LONDON MATHEMATICAL SOCIETY LECTURE NOTE SERIES

1. General cohomology theory and K-theory, P.HILTON
4. Algebraic topology: a student's guide, J.F.ADAMS
5. Commutative algebra, J.T.KNIGHT
8. Integration and harmonic analysis on compact groups, R.E.EDWARDS
9. Elliptic functions and elliptic curves, P.DU VAL
10. Numerical ranges II, F.F.BONSALL & J.DUNCAN
11. New developments in topology, G.SEGAL (ed.)
12. Symposium on complex analysis, Canterbury, 1973, J.CLUNIE & W.K.HAYMAN (eds.)
13. Combinatorics: Proceedings of the British combinatorial conference 1973, T.P.McDONOUGH & V.C.MAVRON (eds.)
14. Analytic theory of abelian varieties, H.P.F.SWINNERTON-DYER
15. An introduction to topological groups, P.J.HIGGINS
16. Topics in finite groups, T.M.GAGEN
17. Differentiable germs and catastrophes, Th.BROCKER & L.LANDER
18. A geometric approach to homology theory, S.BUONCRISTIANO, C.P.ROURKE & B.J.SANDERSON
20. Sheaf theory, B.R.TENNISON
21. Automatic continuity of linear operators, A.M.SINCLAIR
23. Parallelisms of complete designs, P.J.CAMERON
24. The topology of Stiefel manifolds, I.M.JAMES
25. Lie groups and compact groups, J.F.PRICE
26. Transformation groups: Proceedings of the conference in the University of Newcastle upon Tyne, August 1976, C.KOSNIOWSKI
27. Skew field constructions, P.M.COHN
28. Brownian motion, Hardy spaces and bounded mean oscillation, K.E.PETERSEN
29. Pontryagin duality and the structure of locally compact abelian groups, S.A.MORRIS
30. Interaction models, N.L.BIGGS
31. Continuous crossed products and type III von Neumann algebras, A.VAN DAELE
32. Uniform algebras and Jensen measures, T.W.GAMELIN
33. Permutation groups and combinatorial structures, N.L.BIGGS & A.T.WHITE
34. Representation theory of Lie groups, M.F.ATIYAH et al.
35. Trace ideals and their applications, B.SIMON
36. Homological group theory, C.T.C.WALL (ed.)
37. Partially ordered rings and semi-algebraic geometry, G.W.BRUMFIEL
38. Surveys in combinatorics, B.BOLLOBAS (ed.)
39. Affine sets and affine groups, D.G.NORTHCOTT
40. Introduction to H_p spaces, P.J.KOOSIS
41. Theory and applications of Hopf bifurcation, B.D.HASSARD, N.D.KAZARINOFF & Y-H.WAN
42. Topics in the theory of group presentations, D.L.JOHNSON
43. Graphs, codes and designs, P.J.CAMERON & J.H.VAN LINT
44. Z/2-homotopy theory, M.C.CRABB
45. Recursion theory: its generalisations and applications, F.R.DRAKE & S.S.WAINER (eds.)
46. p-adic analysis: a short course on recent work, N.KOBLITZ

London Mathematical Society Lecture Note Series. 41

Theory and Applications of Hopf Bifurcation

B.D.HASSARD,N.D.KAZARINOFF and Y.-H.WAN

Department of Mathematics

State University of New York at Buffalo

CAMBRIDGE UNIVERSITY PRESS

CAMBRIDGE

LONDON NEW YORK NEW ROCHELLE

MELBOURNE SYDNEY

Published by the Press Syndicate of the University of Cambridge
The Pitt Building, Trumpington Street, Cambridge CB2 1RP
32 East 57th Street, New York, NY 10022, USA
296 Beaconsfield Parade, Middle Park, Melbourne 3206, Australia

First published 1981

Printed in the United States of America
Printed and bound by BookCrafters, Inc., Chelsea, Michigan

British Library cataloguing in publication data

Hassard, B D
Theory and applications of Hopf bifurcation.
-(London Mathematical Society Lecture Note Series;41
ISSN 0076-0552).
1. Differential equations, Nonlinear
2. Bifurcation theory
I. Title   II. Kazarinoff, N D
III. Wan, Y H   IV. Series
515'.35   QA371   80-49691

ISBN 0 521 23158 2

CONTENTS

PREFACE

We were motivated to write these Notes by our joint belief that it would be useful to scientists in many fields to have a mature and effective version of the Hopf bifurcation algorithm available together with examples of how to apply it, both by hand and by machine.

Our capacities for writing these Notes were much enhanced by colleagues who both stimulated our efforts and gave generously of their knowledge. We thank them all: Jim Boa, Steve Bernfeld, Shui-Nee Chow, Don Cohen, Paul Fife, Jim Greenberg, Jack Hale, Alan Hastings, Philip Holmes, In-Ding Hsü, Bryce McLeod, Milos Marek, Jerry Marsden, Alistair Mees, Piero de Mottoni, Aubrey Poore, John Rinzel, Reinhard Ruppelt, David Sattinger, Luigi Salvadori, Agnes Schneider, Joel Smoller, and Pauline van den Driessche.

We also thank the National Science Foundation for support under Grants MCS-7905790, MCS-7819647, and MCS-770108.

We owe our typists special thanks. The copy you are reading was typed by Gail Berti. Earlier versions were typed by Marie Daniel, Sue Szydlowski, and Lynda Tomasikiewicz. Their work enabled us to improve these Notes significantly.

Buffalo, N. Y., March, 1980

Brian Hassard
Nicholas Kazarinoff
Yieh-Hei Wan

CHAPTER 1. THE HOPF BIFURCATION THEOREM

1. INTRODUCTION

Periodic phenomena or oscillations are observed in many naturally occurring nonconservative systems. The purpose of these Notes is to describe some of the recent developments in the study of such phenomena; namely, those related to what is called Hopf Bifurcation. In this Introduction we shall limit our attention to Hopf Bifurcation for an autonomous system of ordinary differential equations

$$\frac{dx}{dt} = f(x,\nu) \quad , \tag{1.1}$$

where $x \in \mathbb{R}^n$ and ν is a real valued parameter on an interval $\mathcal{J}$.

The crucial hypotheses made are that (1.1) has an isolated stationary point, say at $x = x_*(\nu)$, and that the Jacobian matrix

$$A(\nu) = D_x f(x_*(\nu),\nu) = \left(\frac{\partial f}{\partial x_j}(x_*(\nu),\nu);\ i,j = 1,\ldots,n\right) \tag{1.2}$$

has a pair of complex conjugate eigenvalues λ_1 and λ_2 ,

$$\lambda_1(\nu) = \bar{\lambda}_2(\nu) = \alpha(\nu) + i\omega(\nu) \tag{1.3}$$

such that for some number $\nu = \nu_c \in \mathcal{J}$

$$\omega(\nu_c) = \omega_0 > 0, \ \alpha(\nu_c) = 0, \text{ and } \alpha'(\nu_c) \neq 0 . \qquad (1.4)$$

The number ν_c is called a critical value of ν. If the eigenvalues of $A(\nu_c)$, other than $\pm i\omega_0$, all have strictly negative real parts, the assumption (1.4) means that there is a loss of linear stability of the stationary point $x_*(\nu)$ as ν crosses ν_c. Under certain additional technical assumptions, we will prove in this Chapter that the system has a family of periodic solutions $x = p_\varepsilon(t)$ $(0 < \varepsilon \leq \varepsilon_0)$, where ε measures the amplitude

$$\max_t \| p_\varepsilon(t) - x_*(\nu_c) \|$$

and ε_0 is sufficiently small. This appearance of periodic solutions out of an equilibrium state is called Hopf bifurcation [55]. Most usually, the periodic solutions exist in exactly one of the cases $\nu > \nu_c$, $\nu < \nu_c$, can be written as $x = p(t,\nu)$ (i.e. indexed by ν) and have amplitudes

$$\text{const. } |\nu - \nu_c|^{\frac{1}{2}} + O(|\nu - \nu_c|) .$$

Bifurcation from stationary to periodic solutions also occurs in functional and integro-differential equations and in partial differential equations. In these different situations we define as a Hopf bifurcation any bifurcation from a stationary to periodic solutions that admits an underlying autonomous two-dimensional ordinary differential system as an "essential model" [54]. Our techniques display the essential model as explicitly as is possible, namely in Poincaré normal form.

Our objectives in these Notes are several: (1) to give a proof of the existence of the (Hopf) bifurcating periodic solutions and to derive formulae describing their periods, amplitudes, and stability, (2) to provide numerical evaluation of these Hopf bifurcation formulae, (3) to study Hopf bifurcation for functional and integro-differential equations, (4) to study Hopf bifurcation for reaction-diffusion equations, and

(5) to provide numerous examples of the theory: both mathematically exact executions of the algorithms of the theory in specific examples, and numerical, computer execution of the algorithms in more complex examples. We also present (see Appendix A) for the more mathematically advanced reader an existence theorem for center manifolds which, in our approach, play a crucial role in the mathematical theory of Hopf Bifurcation.

The history of James Watt's centrifugal governor provides an interesting example of Hopf bifurcation. This device was invented by Watt in about 1782 for the purpose of controlling steam engines, which at that time were used mainly to pump water and to raise elevator cages in mines [78]. The idea behind the governor (see Fig. 3.1, p. 150) is as follows. When the engine slows below the desired speed, the rotating balls of the governor drop, which opens a valve and increases the flow of steam to the engine. When the engine starts to operate too fast, the rotating balls rise, which closes the valve and decreases the flow of steam to the engine. The equilibrium angle of the rotating ball assembly is determined by the balance between the force of gravity and the centrifugal force due to the rotation.

Centrifugal governors worked well for roughly a century after their introduction. By 1868 there were about 75,000 steam engines with Watt governors in England alone [78]. However, curious behaviour was observed in a large percentage of them. Depending upon the operating conditions, governors on many of these steam engines would "hunt" for the right operating speed before settling down; others would oscillate, never attaining a constant angular velocity.

The cure for this misbehaviour came in the form of viscous dampers (dashpots), which were added on to the governors. The amount of damping required to eliminate "hunting" was determined empirically.

According to L. S. Pontryagin [93], the first real

understanding of what had occurred was due to I. A. Vyshnegradskii in 1876 [106], who obtained a stability criterion in the form of an inequality involving the physical parameters. This inequality, when satisfied, guarantees stability of that state of motion in which the engine is running at a constant speed, the governor is rotating at a constant angular velocity, and the rotating ball assembly is at the equilibrium angle.

Vyshnegradskii showed that there is a minimum amount of damping which must be present in the system in order to guarantee stability. When the centrifugal governor was first introduced, naturally occurring frictional losses had been sufficient because the engines and governors were relatively small. As time passed, however, steam engines and governors grew larger and the physical parameters changed, with the result that there was no longer enough damping naturally present. This explanation must also have occurred to the unknown engineer(s) who introduced the dashpot to provide additional damping. Vyshnegradskii's stability criterion made the roles of the various physical parameters clear, and it became possible to calculate the amount of damping required for stability as a function of the other parameters.

Earlier, in 1868 J. C. Maxwell [83] carried out a mathematical analysis of governors of various types and derived a linear stability criterion for the equilibrium state, exposing the role of viscosity (friction). The first mathematical discussion of a servomechanism was apparently given by G. B. Airy in 1840 [2] in connection with the regulation of the clockwork for driving equatorially mounted telescopes. An interesting history of the theory of servomechanisms and control has been written by A. G. J. MacFarlane [78].

Vyshnegradskii obtained his stability criterion by means of linear stability analysis of the system of ordinary differential equations expressing the dynamics of the steam engine, centrifugal governor combination. In Chapter 3 we shall study

Pontryagin's version of this system, which is presented in his textbook [93]. Pontryagin shows, by means of Lyapounov's theorem [93, p. 208], that when Vyshnegradskii's stability criterion is satisfied, the stationary solution is not only linearly stable but is also asymptotically stable as a solution of the nonlinear system. The stationary solution represents the state of motion in which the engine is running at the design speed, the governor is rotating at a constant angular velocity and the ball assembly is at the equilibrium angle.

Historically, the loss of stability occurred as a sequence of machines were constructed having different physical parameters. It is more convenient, however, to think of the instability as occurring in a single machine as a single one of the parameters is changed. The most natural parameter to vary is the amount of damping γ. For concreteness, then, we suppose that the viscous damper on the centrifugal governor for a steam engine modelled by Pontryagin's system is slowly going "bad". The time scale over which this change occurs is assumed to be long enough that the damping may be regarded as a constant so far as the dynamics of the system are concerned.

The damping coefficient γ is thus chosen to play the role of the bifurcation parameter. The stability criterion may then be written as

$$\gamma > \gamma_c ,$$

where γ_c is the critical amount of damping, which is a function of the remaining parameters in the system. As γ is decreased past γ_c, linear stability analysis shows that the loss of stability is due to a complex conjugate pair of eigenvalues of the Jacobian matrix (the coefficient matrix of the linearized system). This system is of order 3; and for γ near γ_c, the single real eigenvalue is negative. The loss of stability at $\gamma = \gamma_c$ thus fits within the framework of Hopf's Bifurcation Theorem [55], according to which periodic solutions appear in addition to the stationary solution. These periodic solutions represent states of motion of the system in which the centrifugal governor rotates at constant angular velocity, except for a small periodic fluctuation in angular velocity.

Hopf's theorem by itself, however, does not predict whether the periodic solutions represent states of motion which can be physically observed as long-time behavior of the system. The periodic solutions may themselves be unstable, hence not observable except under special circumstances. In fact, Hopf's theorem is not even precise about predicting for what values of the damping γ the periodic solutions exist. Rather his theorem offers a three way choice. The solutions exist, for γ near γ_c, in exactly one of the cases

$$\text{(i)}\quad \gamma > \gamma_c \qquad \text{(ii)}\quad \gamma = \gamma_c \qquad \text{(iii)}\quad \gamma < \gamma_c .$$

It was shown in [45] (and we shall see in Chapter 3) that cases (i) and (iii) do indeed arise in the case of the centrifugal governor. Which one occurs depends upon the parameters of the system. It is an open question whether or not case (ii) may arise.

Stability of the periodic solutions is closely related to which case applies. If case (i) applies, then the periodic solutions are unstable. If case (iii) applies, the periodic solutions are stable. Note that in case (i) the range of γ is such that the family of unstable periodic solutions coexists with the family of stable stationary solutions. In case (iii) the families of stable periodic solutions and unstable stationary solutions coexist.

The technical concepts of stability for periodic solutions are slightly more difficult than those for stationary solutions. Although we shall summarize the most relevant material in Section 4, the reader unfamiliar with the definitions and theorems should refer to one or more of the books [13, Chapter 9, 51, 93] at the undergraduate level and [20, 39, 43] at the graduate level for a thorough background.

When case (iii) applies, the periodic solutions are both linearly stable and orbitally asymptotically stable as solutions of the nonlinear system. Physically, case (iii) is the easiest to understand. As the damping γ is decreased past $\gamma = \gamma_c$, the stationary solution becomes unstable; but a stable periodic solution appears and takes its place as the long-term behavior of the system. When γ is slightly less than γ_c, for almost all initial conditions in a neighborhood of the unstable stationary solution, the system will settle into a controlled state of motion, with the speed varying periodically faster and slower then the design speed.

The situation when case (i) applies is much less pleasant. As the damping γ is decreased toward γ_c, for γ slightly greater than γ_c there are unstable periodic solutions coexisting with the stationary state. These unstable solutions represent a mechanism whereby certain small but finite amplitude perturbations from the stationary solution may grow. For γ near γ_c but $\gamma > \gamma_c$, even though the stationary state is both linearly stable and asymptotically stable, it is physically unstable. In the dynamic simulations we have run the growth in amplitude of the perturbations is limited only by the physical constraints. The motion of the system is essentially uncontrolled. For γ slightly less than γ_c, the unstable periodic solution ceases to exist, but uncontrolled growth is still observed.

There are thus good reasons for wanting to decide which of cases (i), (ii) or (iii) applies. We shall show in Chapter 3, Example 2 that for Pontryagin's steam-engine-centrifugal-governor system, if the governor is designed so the equilibrium angle of the ball assembly always exceeds 39.3°, then whenever a loss of stability occurs, case (iii) applies.

Our results for the steam-engine-centrifugal-governor system come rather late, centrifugal governors having been operated successfully at high angles for generations. However, our techniques of analysis also apply to systems of more recent interest. For example, it is generally desired to operate servomechanisms close to the edge of instability so as to attain high slew rates. In such a situation it is desirable to design so that if a loss of stability by the mechanism of Hopf Bifurcation should occur, it is a bifurcation to stable periodic solutions. That way, some degree of control is retained, which in many cases should be sufficient.

Much stimulus for these Notes was provided by the Hodgkin-Huxley model for nerve conduction. The current-clamped Hodgkin-Huxley system displays two distinct Hopf bifurcations as the current stimulus is increased past two critical values I_1 and I_2. Only recently has this behavior been fully explained [44]: unstable periodic solutions arise for $I \approx I_1$, $I < I_1$; and stable periodic solutions arise for $I \approx I_2$, $I < I_2$. The unstable periodic solutions represent a transition mechanism rather than long term behaviour of the nerve axon system; see Chapter 3, Example 4 for details.

The following simple mechanical system illustrates Hopf bifurcation to unstable periodic solutions which represent such a transition mechanism. Similar systems may be found in Andronov, Vitt, and Chaikin [4]. We thank Andres Soom (Mechanical Engineering Department, SUNY at Buffalo) for suggesting the moving belt.

Consider a mass m on a belt that moves with constant velocity ν. We suppose that the mass is tethered to a spring and viscous damper (dashpot) arrangement as shown in Figure 1.1.

We let $x = 0$ denote the position of the mass when the spring has its natural length, and we let $x > 0$ denote displacements to the right. Suppose that the forces due to the spring and dashpot are $-kx$ and $-c\dot{x}$, respectively, where k and c are positive constants. When not stuck to the belt, the

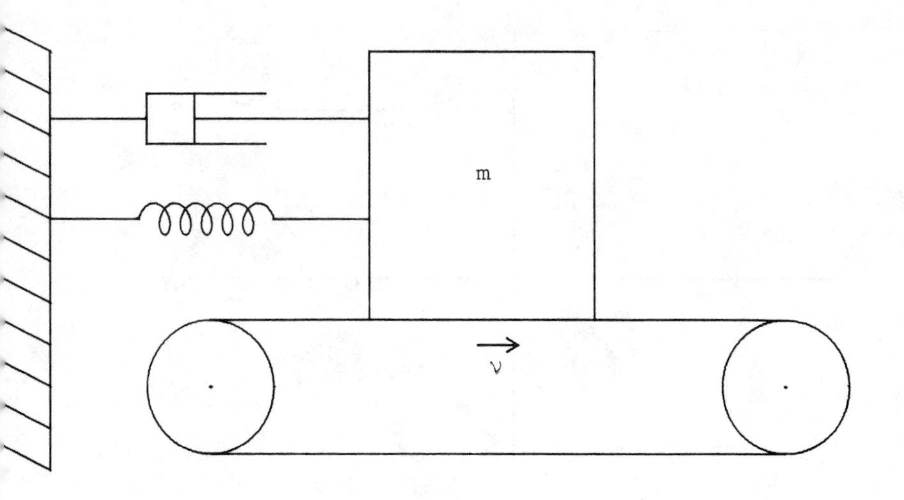

Figure 1.1.

mass m moves according to the law

$$m\ddot{x} = \mathfrak{F}(\nu - \dot{x}) - kx - c\dot{x} , \qquad (1.5)$$

where $\mathfrak{F}(\nu)$ is the frictional force that would be exerted by the belt with velocity ν upon the stationary mass. We assume a Coulomb-type frictional force with the qualitative properties shown in Figure 1.2. A force law of this type combines the effects of both sliding and static friction. It is quite realistic, except at the origin; namely, when $\dot{x} = \nu$ in equation (1.5). If $\dot{x} = \nu$, the mass sticks to the belt, and $\ddot{x} = 0$ until such time as $|kx + c\nu|$ exceeds $\mathfrak{F}_{max} = \sup_{\nu} \mathfrak{F}(\nu)$. Then equation (1.5) again applies. Our Hopf bifurcation analysis will be local; so $\dot{x} \approx 0$, and the case $\dot{x} = \nu$ is outside of its scope.

The following hypotheses are based upon Figure 1.2. We assume there exists a $\nu_0 > 0$ such that

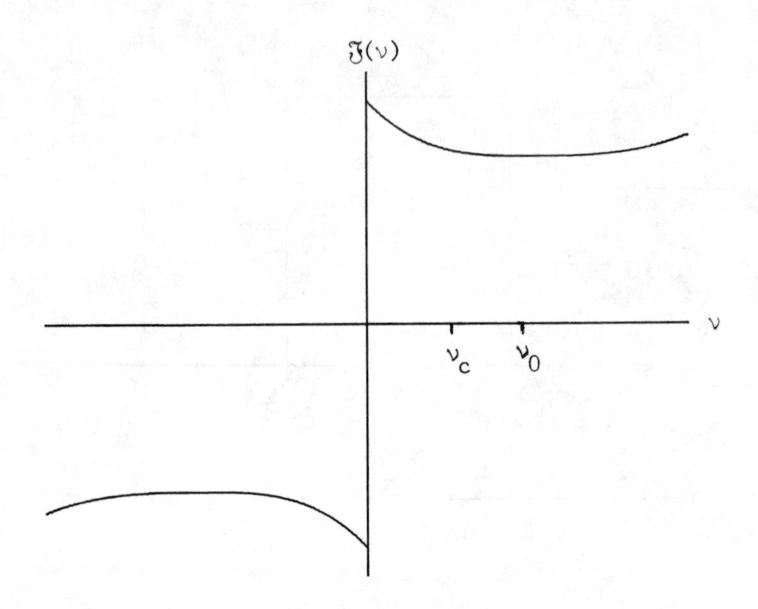

Figure 1.2. Coulomb's frictional force

(i) $\mathfrak{F}'(\nu) < 0$ for $0 < \nu < \nu_0$,

(ii) $\mathfrak{F}'(\nu)$ is increasing for $0 < \nu < \nu_0$,

(iii) $\mathfrak{F}'(\nu_0) = 0$ and $\mathfrak{F}'(\nu) \geq 0$ for $\nu > \nu_0$.

Written as a first order system of the form (1.1), (1.5) becomes

$$\begin{aligned} \dot{x}_1 &= 0 \cdot x_1 + x_2 &&\equiv f^{(1)}(x,\nu) \\ \dot{x}_2 &= -\frac{k}{m} x_1 - \frac{c}{m} x_2 + \frac{1}{m} \mathfrak{F}(\nu - x_2) &&\equiv f^{(2)}(x,\nu) \end{aligned} \tag{1.6}$$

For each $\nu > 0$, (1.6) has the unique equilibrium point

$$x_*(\nu) = \left(\frac{\mathfrak{F}(\nu)}{k} , 0 \right)^T .$$

The Jacobian matrix $D_x f = (\partial f^{(i)}/\partial x_j)$ $(i,j = 1,2)$ at $x_*(\nu)$ is

$$\begin{pmatrix} 0 & 1 \\ -\frac{k}{m} & -\frac{1}{m}\,[c + \mathfrak{F}'(\nu)] \end{pmatrix},$$

which has eigenvalues

$$\lambda_{1,2}(\nu) = \frac{1}{2m}\,\{-[c + \mathfrak{F}'(\nu)] \pm \{[c + \mathfrak{F}'(\nu)]^2 - 4km\}^{\frac{1}{2}}\} .$$

For $\nu \geq \nu_0$, $\mathfrak{F}'(\nu) \geq 0$, Re $\lambda_{1,2}(\nu) < 0$ and $x_*(\nu)$ is a (linearly) stable point of equilibrium. We now make the further assumption

$$\text{(iv)} \quad c < \sup_{0<\nu<\nu_0} |\mathfrak{F}'(\nu)| .$$

This assumption implies the existence of points ν_1 and ν_c $(0 < \nu_c < \nu_1 \leq \nu_0)$ such that for $0 < \nu \leq \nu_1$, $\lambda_1(\nu)$ and $\lambda_2(\nu)$ are complex conjugates and such that Re $\lambda_1(\nu_c) = 0$. Our final assumption is

$$\text{(v)} \quad \mathfrak{F}''(\nu_c) > 0, \; \mathfrak{F}'''(\nu_c) < 0 .$$

For ν near to ν_c

$$\alpha(\nu) \equiv \text{Re}\,\lambda_1(\nu) = -[c + \mathfrak{F}'(\nu)]/(2m)$$

$$\omega(\nu) \equiv \text{Im}\,\lambda_1(\nu) = [\frac{k}{m} - \alpha^2(\nu)]^{\frac{1}{2}} .$$

Then

$$\alpha'(\nu_c) < 0 ,$$

so that the stationary point $x_*(\nu)$ becomes unstable as the belt velocity ν is decreased below ν_c.

Thus the hypotheses for Hopf bifurcation are fulfilled by

this model. We shall show in Chapter 2, Example 1, that the bifurcating solutions exist for $\nu > \nu_c$ and are unstable. These unstable periodic solutions have the form

$$x(t) = \frac{\mathfrak{F}(\nu_c)}{k} + \left(\frac{8m\,\mathfrak{F}''(\nu_c)(\nu - \nu_c)}{k[-\mathfrak{F}'''(\nu_c)]} \right)^{\frac{1}{2}} \cos(2\pi \frac{t}{T}) + O(\nu - \nu_c) \ , \qquad (1.7)$$

where

$$T = \frac{2\pi}{\omega_0} \left(1 - \frac{[\mathfrak{F}''(\nu_c)]^2}{3m^2 \omega_0^2 \mathfrak{F}'''(\nu_c)} (\nu - \nu_c) + O[(\nu - \nu_c)^2] \right) . \qquad (1.8)$$

Since these solutions are unstable, they will not be observed as a long-time behavior of the system. However, if the law of friction could be changed so that $\mathfrak{F}'''(\nu_c) > 0$, stable oscillations would be observed for ν slightly smaller than ν_c .

Unstable periodic solutions still have physical meaning, however; and their presence may be observed as follows. Set up the mass-spring-belt system using a dashpot with a large damping coefficient c . For different constant speeds ν of the belt, record the equilibrium position of the mass m . These measurements determine the force law $\mathfrak{F}(\nu)$. Next replace the dashpot with one having damping coefficient c small enough that the equilibrium position is stable for large ν but unstable for small ν . Roughly locate the value $\nu = \nu_c > 0$ that divides the regions of stability and instability. For several values of ν slightly greater than ν_c , determine the range of initial positions x_0 such that when the mass is released from x_0 with zero velocity, the subsequent motion damps down to the equilibrium position. For initial positions outside this range, the motion will become a large amplitude relaxation oscillation. But for precisely two initial positions x_0 at the limits of this range, it is possible (in principle, if not in practice) to obtain a periodic motion that neither grows nor decays. This is the unstable periodic solution. Finally, plot the limits of the range of initial conditions found to produce decaying motions against ν . The figure found should be approximated by the parabola with equation

$$\frac{8m\,\mathfrak{F}''(\nu_c)}{k[-\mathfrak{F}'''(\nu_c)]}(\nu - \nu_c) = \left(x_0 - \frac{\mathfrak{F}(\nu_c)}{k}\right)^2 . \tag{1.9}$$

For readers who seek advice about the background needed to read the remainder of these Notes we recommend a number of books. At the undergraduate level we again suggest the books [13, 51, 93] by Boyce and DiPrima, Hirsch-Smale, and Pontryagin as well as [5] by Arnol'd. These provide prerequisites for most of Chapters 1 and 2. As preparation for Chapter 3 we suggest the reader select among the numerical analysis books by Conte and deBoor [21], Dahlquist and Björck [27], Ortega and Rheinboldt [89], Wilkinson [109, 110], and Young and Gregory [112] in addition. Background for the more advanced mathematics underlying Chapters 4 and 5 may be found in [20, 24, 34, 36, 38, 39, 40, 43, 49, and 85].

2. THE HOPF BIFURCATION THEOREM

In this section we give alternative versions of E. Hopf's theorem on bifurcation of periodic solutions of ordinary differential equations:

$$\frac{dX}{dt} = F(X,\mu) \ , \tag{2.1}$$

where now the stationary point is $X = 0 \in \mathbb{R}^n$ and the critical value of the bifurcation parameter μ is 0 ; that is, we have defined

$$X = x - x_*(\nu) \quad \text{and} \quad \mu = \nu - \nu_c \ .$$

In terms of the function f of (1.1), the right-hand side of (2.1) is

$$f(X + x_*(\nu_c + \mu), \nu_c + \mu) \ .$$

We say that (2.1) has a family of periodic solutions, indexed by a parameter $\varepsilon \in (0,\varepsilon_0)$ and bifurcating from $X = 0$, if the orbits of the periodic solutions, one for each $\varepsilon \in (0,\varepsilon_0)$, have amplitudes that are $o(1)$ as $\varepsilon \downarrow 0$.

We shall state three versions of the theorem. The first version is analytic; the second is nonanalytic; the third version uses the second to strengthen the uniqueness and analyticity conclusions of the first and describes the family of periodic solutions directly in terms of μ .

For the proof of Theorem I we refer the reader to the annotated translation by L. N. Howard and N. Kopell of Hopf's original paper: [81, pp. 163-205]. We shall prove Theorem III in this Section. Our proof of Theorem II occupies the remainder of this Chapter.

<u>Theorem</u> I (E. Hopf).

<u>If</u>

(1) $F(0,\mu) = 0$ *for* μ *in an open interval containing* 0 , *and* $0 \in R^n$ *is an isolated stationary point of* F ,

(2) F *is analytic in* X *and* μ *in a neighborhood of* $(0,0)$ *in* $R^n \times R^1$,

(3) $A(\mu) = D_X F(0,\mu)$ *has a pair of complex conjugate eigenvalues* λ *and* $\bar{\lambda}$ *such that*

$$\lambda(\mu) = \alpha(\mu) + i\omega(\mu) , \tag{2.2}$$

where

$$\omega(0) = \omega_0 > 0, \ \alpha(0) = 0, \ \alpha'(0) \neq 0 , \tag{2.3}$$

(4) *the remaining* $n - 2$ *eigenvalues of* $A(0)$ *have strictly negative real parts*,

then the system (2.1) *has a family of periodic solutions*: *there is an* $\epsilon_H > 0$ *and an analytic function*

$$\mu^H(\epsilon) = \Sigma_2^\infty \mu_i^H \epsilon^i \qquad (0 < \epsilon < \epsilon_H) \tag{2.4}$$

such that for each $\epsilon \in (0,\epsilon_H)$ *there exists a periodic solution* $p_\epsilon(t)$ *occurring for* $\mu = \mu^H(\epsilon)$. *If* $\mu^H(\epsilon)$ *is not identically zero, the first nonvanishing coefficient* μ_i^H *has an even subscript, and there is an* $\epsilon_1 \in (0, \epsilon_H]$ *such that* $\mu^H(\epsilon)$ *is either strictly positive or strictly negative for* $\epsilon \in (0, \epsilon_1)$. *For each* $L > 2\pi/\omega_0$ *there is a neighborhood* $\mathfrak{n}$ *of* $X = 0$ *and an open interval* $\mathcal{J}$ *containing* 0 *such that for any* $\mu \in \mathcal{J}$ *the only nonconstant periodic solutions of* (2.1) *with periods less than* L *which lie in* $\mathfrak{n}$ *are members of the family* $p_\epsilon(t)$ *for values of* ϵ *satisfying* $\mu^H(\epsilon) = \mu$, $\epsilon \in (0,\epsilon_H)$. *The period* $T^H(\epsilon)$ *of* $p_\epsilon(t)$ *is an analytic function*

$$T^H(\epsilon) = \frac{2\pi}{\omega_0} [1 + \Sigma_2^\infty \tau_i^H \epsilon^i] \qquad (0 < \epsilon < \epsilon_H) . \tag{2.5}$$

Exactly two of the Floquet exponents [93, pp. 144-149 or 20, pp. 78-80, 321-327] of $p_\varepsilon(t)$ approach 0 as $\varepsilon \downarrow 0$. One is 0 for $\varepsilon \in (0, \varepsilon_H)$, and the other is an analytic function

$$\beta^H(\varepsilon) = \Sigma_2^\infty \beta_i^H \varepsilon^i \qquad (0 < \varepsilon < \varepsilon_H) . \tag{2.6}$$

The periodic solution $p_\varepsilon(t)$ is orbitally asymptotically stable with asymptotic phase if $\beta^H(\varepsilon) < 0$ but is unstable if $\beta^H(\varepsilon) > 0$.

Theorem II (C^L-Hopf Bifurcation).

If

(1) $F(0,\mu) = 0$ for μ in an open interval containing 0, and $0 \in \mathbb{R}^n$ is an isolated stationary point of F,

(2) All partial derivatives of the components F^ℓ of the vector F of orders $\leq L + 2$ $(L \geq 2)$ exist and are continuous in X and μ in a neighborhood of $(0,0)$ in $\mathbb{R}^n \times \mathbb{R}^1$,

(3) $A(\mu) = D_X F(0,\mu)$ has a pair of complex conjugate eigenvalues λ and $\bar{\lambda}$ such that

$$\lambda(\mu) = \alpha(\mu) + i\omega(\mu) ,$$

where

$$\omega(0) = \omega_0 > 0, \ \alpha(0) = 0, \ \alpha'(0) \neq 0 ,$$

(4) the remaining $n - 2$ eigenvalues of $A(0)$ have strictly negative real parts,

then the system (2.1) has a family of periodic solutions: there exist an $\varepsilon_P > 0$ and a C^{L+1}-function $\mu^P(\varepsilon)$,

$$\mu^P(\varepsilon) = \Sigma_1^{[\frac{L}{2}]} \mu_{2i}^P \varepsilon^{2i} + O(\varepsilon^{L+1}) \qquad (0 < \varepsilon < \varepsilon_P) \tag{2.7a}$$

such that for each $\varepsilon \in (0, \varepsilon_P)$ there exists a periodic solution

$p_\varepsilon(t)$, *occurring for* $\mu = \mu^P(\varepsilon)$. *There is a neighborhood* $\mathcal{N}$ *of* $X = 0$ *and an open interval* $\mathcal{I}$ *containing* 0 *such that for any* $\mu \in \mathcal{I}$ *the only nonconstant periodic solutions of* (2.1) *that lie in* $\mathcal{N}$ *are members of the family* $p_\varepsilon(t)$ *for values of* ε *satisfying* $\mu^P(\varepsilon) = \mu$, $\varepsilon \in (0, \varepsilon_P)$. *The period* $T^P(\varepsilon)$ *of* $p_\varepsilon(t)$ *is a* C^{L+1}*-function*

$$T^P(\varepsilon) = \frac{2\pi}{\omega_0}\left[1 + \sum_1^{[\frac{L}{2}]} \tau_{2j}^P \varepsilon^{2i}\right] + O(\varepsilon^{L+1}) \qquad (0 < \varepsilon < \varepsilon_P) . \tag{2.7b}$$

Exactly two of the Floquet exponents of $p_\varepsilon(t)$ *approach* 0 *as* $\varepsilon \downarrow 0$. *One is* 0 *for* $\varepsilon \in (0, \varepsilon_P)$, *and the other is a* C^{L+1} *function*

$$\beta^P(\varepsilon) = \sum_1^{[\frac{L}{2}]} \beta_{2i}^P \varepsilon^{2i} + O(\varepsilon^{L+1}) \qquad (0 < \varepsilon < \varepsilon_P) . \tag{2.7c}$$

The periodic solution $p_\varepsilon(t)$ *is orbitally asymptotically stable with asymptotic phase if* $\beta^P(\varepsilon) < 0$ *but is unstable if* $\beta^P(\varepsilon) > 0$. *If there exists a first nonvanishing coefficient* μ_{2K}^P $(1 \le K \le [L/2])$, *then there is an* $\varepsilon_1 \in (0, \varepsilon_p]$ *such that the open interval*

$$\mathcal{I}_1 = \{\mu \mid 0 < \mu/\mu_{2K}^P < \mu^P(\varepsilon_1)/\mu_{2K}^P\} \tag{2.8}$$

has the following properties. For any μ *in* $\mathcal{I}_1$ *there is a unique* $\varepsilon \in (0, \varepsilon_1)$ *for which* $\mu^P(\varepsilon) = \mu$. *Hence the family of periodic solutions* $p_\varepsilon(t)$ $(0 < \varepsilon < \varepsilon_1)$ *may be parameterized as* $p(t;\mu)$ $(\mu \in \mathcal{I}_1)$. *For* $\mu \in \mathcal{I}_1$, *the period* $T(\mu)$ *and Floquet exponent* $\beta(\mu)$ *are* C^L*-functions of* $|\mu|^{1/K}$. *The first non-vanishing coefficient* β_{2K}^P *is given by*

$$\beta_{2K}^P = -2\alpha'(0)\mu_{2K}^P \tag{2.9a}$$

and

$$\operatorname{sgn} \beta(\mu) = \operatorname{sgn} \beta_{2K}^{P} \qquad (\mu \in \mathcal{J}_1) . \tag{2.9b}$$

<u>Thus the members</u> $p(t;\mu)$ $(\mu \in \mathcal{J}_1)$ <u>of the family of periodic solutions are orbitally asymptotically stable with asymptotic phase if</u> $\beta_{2K}^{P} < 0$ <u>and are unstable if</u> $\beta_{2K}^{P} > 0$.

<u>Theorem</u> III.

<u>Suppose that the hypotheses of Theorem</u> I <u>are satisfied. Also assume there exists a first nonvanishing coefficient</u> μ_{2K}^{P} <u>in the expansion</u> (2.7a). <u>Then there exist a neighborhood</u> $\tilde{\eta}$ <u>of</u> $X = 0$ <u>and a number</u> $\epsilon_1 > 0$ <u>such that for each</u> $\mu \in \mathcal{J}_1$, <u>where</u> $\mathcal{J}_1$ <u>is given by</u> (2.8), <u>there exists exactly one nonconstant periodic solution</u> $p(t;\mu)$ <u>of</u> (2.1) <u>which lies in</u> $\tilde{\eta}$. <u>The period</u> $T(\mu)$ <u>is an analytic function of</u> $|\mu|^{1/K}$. <u>Exactly two of the Floquet exponents of</u> $p(t;\mu)$ <u>approach</u> 0 <u>as</u> $\mu \to 0$, $\mu \in \mathcal{J}_1$. <u>One is</u> 0 <u>for</u> $\mu \in \mathcal{J}_1$, <u>and the other is a real valued function</u> $\beta(\mu)$, <u>analytic in</u> $|\mu|^{1/K}$. <u>For each</u> μ <u>in the interval</u> $\mathcal{J}_1$,

$$\operatorname{sgn} \beta(\mu) = \operatorname{sgn} \beta_{2K}^{P} ,$$

<u>where</u> β_{2K}^{P} <u>is described by</u> (2.9a). <u>The periodic solutions</u> $p(t;\mu)$ <u>are orbitally asymptotically stable with asymptotic phase if</u> $\beta_{2K}^{P} < 0$ <u>and are unstable if</u> $\beta_{2K}^{P} > 0$.

We amplify these theorems in the following Remarks.

<u>Remark</u> 1. If, for all sufficiently small positive ϵ , the periodic solutions $p_\epsilon(t)$ in Theorems I, II or III exist for $\mu > 0$ (resp. $\mu = 0$, $\mu < 0$), the <u>direction of bifurcation</u> is defined to be +1 (resp. 0, -1) . If there exists a first nonvanishing coefficient μ_{2K}^{P} (or μ_{2J}^{H}) , then its sign determines the direction of bifurcation. In applications, it is more

difficult to find the direction of bifurcation than it is to determine that a Hopf bifurcation occurs. In the remaining sections of this Chapter we shall derive and describe Bifurca-tion Formulae that enable one to evaluate

$$\mu_2^P, \mu_4^P, \tau_2^P, \tau_4^P, \beta_2^P, \text{ and } \beta_4^P .$$

If $\mu_2^P \neq 0$, or if $\mu_2^P = 0$ and $\mu_4^P \neq 0$, then the direction of bifurcation is explicitly determined from these Bifurcation Formulae.

The terms supercritical and subcritical also appear in the literature. If $\alpha'(0) > 0$, periodic solutions which exist for $\mu > 0$ (resp. $\mu < 0$) are termed supercritical (resp. subcritical). If $\alpha'(0) < 0$, however, there is disagreement about the definitions of these terms, and we suggest that the reader proceed cautiously when encountering this situation.

Remark 2. Theorem I is a restatement of E. Hopf's original version [81, pp. 163-205 and 55]. A defect of the original remains in Theorem I: namely, the possible presence of odd powers of ε in the expansions of $\mu^H(\varepsilon)$, $\beta^H(\varepsilon)$, and $T^H(\varepsilon)$. In consequence, Hopf's Theorem (Theorem I) does not provide the full analyticity results possible for the functions $T(\mu)$ and $\beta(\mu)$. If the Hopf expansion of $\mu^H(\varepsilon)$ has a first non-vanishing coefficient μ_{2J}^H , Theorem I only proves analyticity of $T(\mu)$ and $\beta(\mu)$ in $|\mu|^{1/2J}$. In this same case, we claim that the first nonvanishing coefficient in the expansion (2.7a) of $\mu^P(\varepsilon)$ occurs for the same index $2K = 2J$. (The proof involves a comparison of the role of ε in the derivations of Theorems I and II.) Thus whenever $\mu^H(\varepsilon)$ does not vanish identically, i.e. has a first nonvanishing coefficient μ_{2J}^H , Theorem III shows that $T(\mu)$ and $\beta(\mu)$ are actually analytic in $|\mu|^{1/J}$. In particular, if $\mu_2 \neq 0$ (it has been shown that

$\mu_2 = \mu_2^H = \mu_2^P$ [46]) both $T(\mu)$ and $\beta(\mu)$ are analytic in μ.

Another consequence of the odd powers of ε in Theorem I is the appearance in the literature of approximations of the form

$$\mu(\varepsilon) = \mu_2\varepsilon^2 + \mu_3\varepsilon^3 + O(\varepsilon^4) ,$$

$$T(\varepsilon) = \frac{2\pi}{\omega_0}(1 + \tau_2\varepsilon^2 + \tau_3\varepsilon^3) + O(\varepsilon^4)$$

$$\beta(\varepsilon) = \beta_2\varepsilon^2 + \beta_3\varepsilon^3 + O(\varepsilon^4) ,$$

in which some of μ_3, τ_3, β_3 are nonzero. When encountering such approximations, perform the following substitution. Let $\varepsilon = \tilde{\varepsilon} + \gamma\tilde{\varepsilon}^2$, and form

$$\mu = \mu_2\tilde{\varepsilon}^2 + \tilde{\mu}_3\tilde{\varepsilon}^3 + O(\tilde{\varepsilon}^4) ,$$

$$T = \frac{2\pi}{\omega_0}(1 + \tau_2\tilde{\varepsilon}^2 + \tilde{\tau}_3\tilde{\varepsilon}^3) + O(\tilde{\varepsilon}^4) ,$$

$$\beta = \beta_2\tilde{\varepsilon}^2 + \tilde{\beta}_3\tilde{\varepsilon}^3 + O(\tilde{\varepsilon}^4) .$$

Then, for some choice of γ, all of $\tilde{\mu}_3$, $\tilde{\tau}_3$, and $\tilde{\beta}_3$ will vanish simultaneously. This demonstrates that the appearance of μ_3, τ_3, and β_3 was caused by an unfortunate choice of algorithm for computing the bifurcating solutions.

Remark 3. The case where some (or all) of the remaining $n - 2$ eigenvalues of $A(0)$ (cf. Hypothesis (4) above) have strictly positive real parts for all $\mu \in \mathcal{I}$ is easily handled. The periodic solutions that arise are, however, unstable because the eigenvalues with positive real parts give rise to characteristic (Floquet) exponents with positive real parts.

A variety of interesting phenomena, including bifurcation to tori, may occur if some (or all) of the remaining $n - 2$

eigenvalues are on the imaginary axis when $\mu = 0$; see [62, pp. 67-69; 81, pp. 206-218], for example.

Remark 4. The system

$$\dot{r} = \mu r - r^{2\ell} \sin(\tfrac{1}{r}) ,$$
$$\dot{\theta} = 1 ,$$

(in polar coordinates) or

$$\begin{aligned} \dot{x}_1 &= \mu x_1 - x_2 - x_1 r^{2\ell-1} \sin(\tfrac{1}{r}) \\ \dot{x}_2 &= x_1 + \mu x_2 - x_2 r^{2\ell-1} \sin(\tfrac{1}{r}) \end{aligned} \qquad (r^2 = x_1^2 + x_2^2) ,$$

(in Cartesian coordinates), where ℓ is a positive integer, has finitely many periodic solutions for each μ in a full neighborhood of $0 \in \mathbb{R}^1$ and infinitely many for $\mu = 0$. In this example,

$$\alpha(\mu) = \mu , \ \omega(\mu) = 1 , \text{ and } \ \mu(\varepsilon) = \varepsilon^{2\ell-1} \sin(\tfrac{1}{\varepsilon}) .$$

The orbits of the periodic solutions of the system are circles with radii determined from the equation

$$\mu = r^{2\ell-1} \sin(\tfrac{1}{r}) .$$

Theorem I does not apply to this example; but, if $\ell \geq 5$, Theorem II does apply. There is one periodic solution for each ε , but there are different numbers of periodic solutions for different values of μ .

Remark 5. In Theorem II, if $F \in C^{\infty}$ jointly in its arguments, $\mu^{P}_{2K} = 0$ for $k = 1,2,...,$ and $0 \in \mathbb{R}^n$ is attracting for $\mu = 0$, one cannot conclude stability; see Chafee [15] and Negrini-Salvadori [88].

Remark 6. The difference between Theorems I and II lies with the conclusions involving uniqueness and analyticity of μ, T, and β. We shall prove Theorem II in Sections 3-6 of this Chapter using the Center Manifold Theorem and a reduction of (2.1) to Poincaré normal form. This approach yields a stronger uniqueness result than that of Theorem I, but this is balanced by a loss of analyticity of μ, T, and β in ε in Theorem II. Clearly, Theorem II can be used with analytic hypotheses to sharpen the uniqueness result of Theorem I. What is not so obvious is that Theorem II also sharpens the analyticity results, with respect to μ, of Theorem I. We do this in the following:

Proof of Theorem III.

Theorem I establishes the existence of analytic functions $\mu^H(\varepsilon)$, $T^H(\varepsilon)$, and $\beta^H(\varepsilon)$ (superscript H for Hopf). By the assumption $\mu^P_{2K} \neq 0$, the bifurcating periodic solutions occur for $\mu \neq 0$; hence, there exists a first nonvanishing coefficient μ^H_{2J} in the power series

$$\mu^H(\varepsilon) = \sum_{j=2}^{\infty} \mu^H_j \varepsilon^j .$$

Therefore, by Lagrange's theorem on reversion of series [23, pp. 123-125], there exists an inverse function $E^H(\mu)$ that is analytic in the variable $|\mu|^{1/2J}$ and is such that

$$E^H(\mu^H(\varepsilon)) = \varepsilon \qquad (0 \le \varepsilon \le \varepsilon_0) ,$$

$$\mu^H(E^H(\mu)) = \mu \qquad (0 \le \mu/\mu^H_{2J} \le \mu^H(\varepsilon_0)/\mu^H_{2J})$$

for some sufficiently small positive ε_0. Theorem II also applies under the hypotheses of Theorem I and establishes the existence of functions $\mu^P(\varepsilon)$, $T^P(\varepsilon)$, $\beta^P(\varepsilon)$ (superscript P for Poincaré) that are C^L in ε for any positive integer L and which have expansions in even powers of ε. For any integer

$I > K$, the function $\mu^P(\epsilon)$ has an expansion

$$\mu^P(\epsilon) = \sum_{k=K}^{I-1} \mu_{2k}^P \epsilon^{2k} + O(\epsilon^{2I}) ;$$

and since $\mu_{2K}^P \neq 0$, there exists an inverse function $E^P(\mu)$ of the form

$$E^P(\mu) = \left[\sum_{k=1}^{M-1} e_k^P |\mu|^{k/K} + O(|\mu|^{M/K}) \right]^{1/2} ,$$

where $M = I + 1 - K$. This function $E^P(\mu)$ has the properties

$$E^P(\mu^P(\epsilon)) = \epsilon \qquad (0 \leq \epsilon \leq \epsilon_1) ,$$

$$\mu^P(E^P(\mu)) = \mu \qquad (0 \leq \mu/\mu_{2K}^P \leq \mu^P(\epsilon_1)/\mu_{2K}^P)$$

for some sufficiently small positive ϵ_1.

Now, since the periodic solutions are unique,

$$\text{sgn}\ \mu_{2K}^P = \text{sgn}\ \mu_{2J}^H ;$$

and, further,

$$T(\mu) = T^P(E^P(\mu)) = T^H(E^H(\mu)) ,$$

$$\beta(\mu) = \beta^P(E^P(\mu)) = \beta^H(E^H(\mu))$$

for all sufficiently small μ having the same sign as μ_{2K}^P.

Since $T^H(\epsilon)$ and $\beta^H(\epsilon)$ are analytic in ϵ and $E^H(\mu)$ is analytic in $|\mu|^{1/2J}$, $T(\mu)$ and $\beta(\mu)$ are analytic in $|\mu|^{1/2J}$. However $T^P(E^P(\mu))$ and $\beta^P(E^P(\mu))$ may be expanded to arbitrary order in powers of $|\mu|^{1/K}$. Thus the only powers of $|\mu|^{1/2J}$ that appear in the power series for $T(\mu)$ and

$\beta(\mu)$ are powers of $|\mu|^{1/K}$. This completes the proof of Theorem III.

Remark 1. In the usual case $\overset{P}{\mu_2} \neq 0$, the period $T(\mu)$ and the Floquet exponent $\beta(\mu)$ are analytic functions of μ, for all sufficiently small μ.

Remark 2. The explicit formula for $\overset{H}{\mu_2}$ derived in [57] is the same as the formula for $\overset{P}{\mu_2}$ derived in [46]. Thus we may set

$$\mu_2 = \overset{P}{\mu_2} = \overset{H}{\mu_2} .$$

In these Notes we shall follow the Center Manifold approach [46, 81], and for convenience we adopt the notations

$$\mu_j = \overset{P}{\mu_j} , \quad \beta_j = \overset{P}{\beta_j} , \quad \tau_j = \overset{P}{\tau_j} \quad (j = 1,2,\ldots) .$$

3. EXISTENCE OF PERIODIC SOLUTIONS AND POINCARÉ NORMAL FORM

Our goal in the remainder of this Chapter is to give a proof of Theorem II. Our proof is divided into three main parts: Part I. 2×2 systems in Poincaré normal form; Part II. The reduction of general 2×2 systems to Poincaré normal form; Part III. Application of the Center Manifold Theorem to reduce general n×n systems to the 2×2 case on the center manifold.

There are more efficient techniques available for proving Theorem II [25, 94] than those we use to produce the apparently roundabout proof given in these *Notes*. We have proceeded as we have both for pedagogical reasons and because we desired to exhibit efficient algorithms for computing the form of the bifurcating periodic solutions, their periods, and their stability. The formulae which are the end results are composed of constituent parts, the different parts corresponding to segments in our proof, and each part being significantly simpler than the whole.

Part I(A). Existence.

We assume we are given a 2×2 system in the following *Poincaré* normal form [6, Chapter 5]:

$$\dot{X} = A(\mu)X + \sum_{j=1}^{[\frac{L}{2}]} B_j(\mu)X|X|^{2j} + O(|X||(X,\mu)|^{L+1}) \qquad (3.1)$$

$$= F(X,\mu) \ ,$$

where

$$A(\mu) = \begin{pmatrix} \alpha(\mu) & -\omega(\mu) \\ \omega(\mu) & \alpha(\mu) \end{pmatrix} \qquad (\lambda(\mu) = \alpha(\mu) + i\omega(\mu)) \ , \qquad (3.2)$$

$$B_j(\mu) = \begin{pmatrix} \operatorname{Re} c_j(\mu) & -\operatorname{Im} c_j(\mu) \\ \operatorname{Im} c_j(\mu) & \operatorname{Re} c_j(\mu) \end{pmatrix} \qquad (1 \le j \le [\tfrac{L}{2}]) \ , \quad (3.3)$$

and $F(X,\mu)$ is jointly C^{L+2} in X and μ.

Equation (3.1) is equivalent to

$$\dot{\xi} = \lambda(\mu)\xi + \Sigma_{j=1}^{[\frac{L}{2}]} c_j(\mu)\xi|\xi|^{2j} + o(|\xi||(\xi,\mu)|^{L+1}) \ , \quad (3.4)$$

where $\xi = X_1 + iX_2$.

<u>Remark</u>. Some material on the uses and history of Poincaré's normal form is given by Arnol'd in [6, Chapters 5 and 6]. Although we have not seen an instance in the work of Poincaré in which he makes explicit use of this normal form, all the ideas necessary for derivation of the form (3.1) or (3.4) are present in Poincaré's work, and we feel that it is appropriate to attach his name to this form; see, for example [92]. We note that Poincaré does introduce what is now often called Birkhoff normal form for Hamiltonian systems [10; pp. 82-85].

Following E. Hopf's method, we let $X = \epsilon y$, and we consider the system

$$\dot{y} = A(\mu)y + \Sigma_{j=1}^{[\frac{L}{2}]} \epsilon^{2j} B_j(\mu)y|y|^{2j} + o(|y||(\epsilon y,\mu)|^{L+1}) \quad (3.5)$$

with initial condition $y(0) = (1,0)^T$, which is $(\epsilon,0)^T$ for (3.1). (The superscript T denotes "transpose.") Note that (3.5) makes sense even for $\epsilon = 0$. Since $F(X,\mu)$ is C^{L+2} jointly in X and μ, the right-hand side of equation (3.5) is C^{L+1} jointly in y, μ, and ϵ. Let $y = y(t,\epsilon,\mu)$ denote the solution of (3.5) that satisfies the given initial condition. By standard theory [43, Chapter 5], for $\varepsilon = 0$ and μ small enough, $y(t, 0, \mu)$ exists sufficiently long to cross the positive y_1-axis for $t = T_0(\mu)$, where

$$T_0(\mu) = \frac{2\pi}{\omega(\mu)} + O(\mu^{L+1}) ;$$

see Figure 1.3.

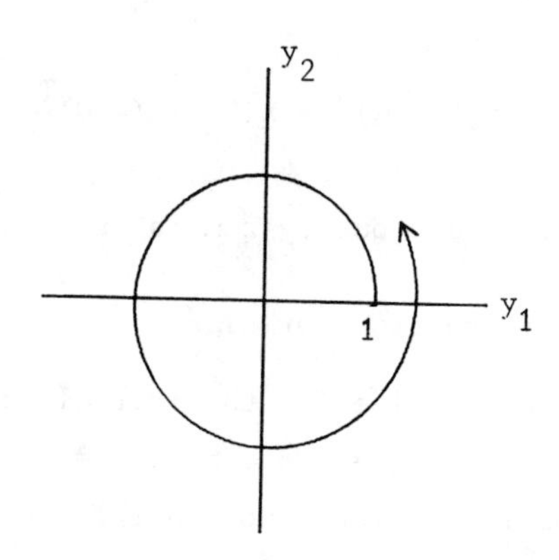

Figure 1.3.

Also, y is C^{L+1} jointly in t, ε, and μ. Now,

$$y(T_0(\mu),0,\mu) = (1,0)^T[e^{2\pi\alpha(\mu)/\omega(\mu)} + O(\mu^{L+1})] ,$$

and from (3.5)

$$\dot{y}(T_0(\mu),0,\mu) = (\alpha(\mu),\omega(\mu))^T e^{2\pi\alpha(\mu)/\omega(\mu)} + O(\mu^{L+1}) .$$

Since $\omega(\mu) > 0$ for μ in a neighborhood of 0, we may apply the Implicit Function Theorem to solve the equation $y_2(t, \varepsilon, \mu) = 0$ for a function $t = T(\varepsilon,\mu)$ with the properties $T(0,\mu) = T_0(\mu)$ and $T \in C^{L+1}$ jointly in ε and μ.

Let

$$I(\varepsilon,\mu) \equiv y_1(T(\varepsilon,\mu), \varepsilon, \mu) . \tag{3.6}$$

Then, for all sufficiently small ε and μ, $I \in C^{L+1}$ jointly in ε and μ. Since $I_\mu(0,0) = 2\pi\alpha'(0)/\omega_0$, we may again apply the Implicit Function Theorem to conclude that for some $\varepsilon_P > 0$ there exists a function $\mu = \mu(\varepsilon)$, C^{L+1} in ε for $\varepsilon \in [0,\varepsilon_P)$, such that $I(\varepsilon,\mu(\varepsilon)) \equiv 1$. We have therefore proved the existence of a family of periodic solutions, one for each $\varepsilon \in (0,\varepsilon_P)$ of

$$\dot{X} = F(X,\mu) \;, \quad X(0) = (\varepsilon,0)^T \;,$$

in the 2x2 case, provided our system is in Poincaré normal form.

Part I(B). Bifurcation Formulae

We next derive formulae for the initial coefficients in the MacLaurin expansions of $\mu = \mu(\varepsilon)$ and $T = T(\varepsilon)$ $(= T(\varepsilon,\mu(\varepsilon)))$. We begin with some motivation. Consider the differential equation

$$\dot{z} = \lambda(\mu)z + z\,\Sigma_1^M\, c_j(\mu)(z\bar{z})^j \;, \tag{3.7}$$

where z is a complex variable, $\lambda(0) = i\omega_0$, $M \geq 1$ is arbitrary, and the $c_j(\mu)$ are complex valued. This canonical equation is in Poincaré normal form. Observe further that (3.7) is rotationally invariant: if z is a solution, then so is $ze^{i\phi}$ for any real number ϕ, and the trajectories of (3.7) are circles with centers at $z = 0$. This simple geometry is reflected in efficient computation of the MacLaurin expansions of $\mu(\varepsilon)$ and $T(\varepsilon)$.

Forming $\bar{z}\,\dot{z} + \dot{\bar{z}}\,z$ from (3.7), we obtain

$$\frac{d}{dt}(z\bar{z}) = 2z\bar{z}\,\{\mathrm{Re}\,\lambda(\mu) + \Sigma_1^M\,\mathrm{Re}\,c_j(\mu)(z\bar{z})^j\} \;. \tag{3.8}$$

The right-hand side of (3.8) is zero if and only if $z = 0$ or

$$\operatorname{Re} \lambda(\mu) + \Sigma_1^M \operatorname{Re} c_j(\mu)(z\bar{z})^j = 0 . \tag{3.9}$$

But if (3.9) holds, then (3.8) implies that $z\bar{z} = \varepsilon^2 \geq 0$, for some $\varepsilon \geq 0$. Setting $z\bar{z} = \varepsilon^2$ and $\mu = \mu(\varepsilon)$ in (3.9) now gives

$$\operatorname{Re} \lambda(\mu(\varepsilon)) + \Sigma_1^M \operatorname{Re} c_j(\mu(\varepsilon))\varepsilon^{2j} = 0 . \tag{3.10}$$

This equation determines the coefficients in the expansion

$$\mu = \Sigma_1^M \mu_j \varepsilon^j + O(\varepsilon^{M+1}) .$$

(In the analysis below of the case $\alpha'(0) \neq 0$, the coefficients $\mu_1, \mu_3, \mu_5, \ldots$ are all shown to vanish, which is a priori obvious from (3.10). However, we shall later consider the case $\alpha'(0) = 0$, $\alpha''(0) \neq 0$, in which case μ_1 is in general nonzero.) Expanding μ in powers of ε in (3.10), we find that

$$\alpha'(0)\Sigma_1^M \mu_j \varepsilon^j + \frac{\alpha''(0)}{2} (\Sigma_1^M \mu_j \varepsilon^j)^2 + \ldots + \operatorname{Re} c_1(0)\varepsilon^2$$

$$+ \operatorname{Re} c_1'(0)[\Sigma_1^M \mu_j \varepsilon^j]\varepsilon^2 + \ldots + \operatorname{Re} c_2(0)\varepsilon^4 + \ldots = 0 . \tag{3.11}$$

At $O(\varepsilon)$, (3.11) implies that $\alpha'(0)\mu_1 = 0$. Thus

$$\mu_1 = 0 , \tag{3.12}$$

since $\alpha'(0) \neq 0$ by hypothesis. Using this result in (3.11), we find that at $O(\varepsilon^2)$

$$\alpha'(0)\mu_2 + \operatorname{Re} c_1(0) = 0 ,$$

$$\mu_2 = - \frac{\operatorname{Re} c_1(0)}{\alpha'(0)} . \tag{3.13}$$

At $O(\epsilon^3)$, (3.11) implies that $\alpha'(0)\mu_3 = 0$. Therefore

$$\mu_3 = 0 \ . \tag{3.14}$$

Using (3.12) and (3.14) in (3.11), we obtain at $O(\epsilon^4)$:

$$\alpha'(0)\mu_4 + \frac{\alpha''(0)}{2}\mu_2^2 + \mu_2 \operatorname{Re} c_1'(0) + \operatorname{Re} c_2(0) = 0 \ ,$$

or

$$\mu_4 = -\frac{1}{\alpha'(0)}\left[\operatorname{Re} c_2(0) + \mu_2 \operatorname{Re} c_1'(0) + \frac{\alpha''(0)}{2}\mu_2^2\right] , \tag{3.15}$$

where μ_2 is given by (3.13).

Given that (3.10) holds, we may rewrite (3.7) as

$$\dot{z} = iz \operatorname{Im}[\lambda(\mu) + \Sigma_1^M c_j(\mu)\epsilon^{2j}] \ . \tag{3.16}$$

Thus

$$z = \epsilon e^{2\pi it/T(\epsilon)} ,$$

where

$$\frac{2\pi}{T(\epsilon)} = \operatorname{Im}[\lambda(\mu(\epsilon)) + \Sigma_1^M c_j(\mu(\epsilon))\epsilon^{2j}] \ . \tag{3.17}$$

From this equation the coefficients in the expansion

$$T(\epsilon) = \frac{2\pi}{\omega_0}\Sigma_0^M \tau_i\epsilon^i + O(\epsilon^{M+1})$$

may be found. Explicitly at $O(1)$, (3.17) yields

$$\tau_0 = 1 \ ; \tag{3.18}$$

and, to higher order,

$$\omega_0(-\Sigma_1^4 \tau_i \varepsilon^i) + \omega_0(\Sigma_1^3 \tau_i \varepsilon^i)^2 + \ldots = \omega'(0)(\mu_2 + \mu_4\varepsilon^2)\varepsilon^2 +$$

$$+ \frac{\omega''(0)}{2} \mu_2^2 \varepsilon^4 + \text{Im } c_1(0)\varepsilon^2 + [\text{Im } c_1'(0)\mu_2 + \text{Im } c_2(0)]\varepsilon^4 + \ldots . \tag{3.19}$$

Hence

$$-\omega_0 \tau_1 = 0 ;$$

or

$$\tau_1 = 0 , \tag{3.20}$$

since $\omega_0 > 0$ by hypothesis. Then at $O(\varepsilon^2)$ (3.19) becomes

$$-\omega_0 \tau_2 = \omega'(0)\mu_2 + \text{Im } c_1(0)$$

$$\tau_2 = \frac{-1}{\omega_0} [\text{Im } c_1(0) + \mu_2\omega'(0)] . \tag{3.21}$$

At $O(\varepsilon^3)$ (3.19) yields

$$-\omega_0 \tau_3 = 0 ,$$

or

$$\tau_3 = 0 . \tag{3.22}$$

Finally, at $O(\varepsilon^4)$ (3.19) becomes

$$-\omega_0 \tau_4 + \omega_0 \tau_2^2 = \omega'(0)\mu_4 + \frac{\omega''(0)}{2} \mu_2^2 + \text{Im } c_1'(0)\mu_2 + \text{Im } c_2(0) ,$$

or

$$\tau_4 = -\frac{1}{\omega_0} [\omega'(0)\mu_4 + \frac{\omega''(0)}{2} \mu_2^2 + \text{Im } c_1'(0)\mu_2 + \text{Im } c_2(0) - \omega_0\tau_2^2], \tag{3.23}$$

where μ_4 is given by (3.15), μ_2 by (3.13) and τ_2 by (3.21).

Remark. If $\alpha'(0) = 0$ but $\alpha''(0) \neq 0$, the algorithm given above for computing the μ_j's and τ_j's can still be used. In this case (3.11) becomes

$$\frac{\alpha''(0)}{2} \{\Sigma_1^M \mu_j \epsilon^j\}^2 + \ldots + \mathrm{Re}\, c_1(0)\epsilon^2 + O(\epsilon^3) = 0 .$$

Thus if $\mathrm{Re}\, c_1(0)$ and $\alpha''(0)$ have opposite signs,

$$\mu_1 = \{-2\mathrm{Re}\, c_1(0)/\alpha''(0)\}^{\frac{1}{2}} ;$$

and there exist bifurcating periodic solutions in a full neighborhood of $\mu = 0$ (see [46]). If, however, $\alpha'(0) = \alpha''(0) = 0$ but $\alpha'''(0) \neq 0$, the above algorithm fails, and fractional powers of ϵ must be introduced in the expansions of $\mu(\epsilon)$ and $T(\epsilon)$.

The computation of exact formulae for the μ_i and τ_i $(i \geq 5)$ can easily be carried further, but we do not compute more of these coefficients as they are not needed in the applications which follow in subsequent chapters.

We shall next show that the bifurcation formulae derived above for systems of the specific form (3.7) actually hold for general 2×2 systems in the Poincaré normal form (3.24) below.

Lemma. *If the Poincaré normal form of* (2.1) *is*

$$\dot{\xi} = \lambda(\mu)\xi + \Sigma_{j=1}^{[\frac{L}{2}]} c_j(\mu)\xi|\xi|^{2j} + O(|\xi||(\xi,\mu)|^{L+1}) \tag{3.24}$$

$$\equiv C(\xi,\bar{\xi},\mu) ,$$

where $C(\xi,\bar{\xi},\mu)$ *is* C^{L+2} *jointly in* ξ, $\bar{\xi}$, μ *in a neighborhood of* $0 \in \mathbb{C} \times \mathbb{C} \times \mathbb{R}^1$, *then the periodic solution of period* $T(\epsilon)$ *such that* $\xi(0,\mu) = \epsilon$ *of* (3.24) *has the form*

$$\xi = \varepsilon \exp[2\pi it/T(\varepsilon)] + O(\varepsilon^{L+2}) , \tag{3.25}$$

where

$$T(\varepsilon) = \frac{2\pi}{\omega_0}[1 + \Sigma_1^L \tau_i \varepsilon^i] + O(\varepsilon^{L+1}) \tag{3.26}$$

and

$$\mu(\varepsilon) = \Sigma_1^L \mu_i \varepsilon^i + O(\varepsilon^{L+1}) , \tag{3.27}$$

and the μ_i *and* τ_i *are again given by* (3.12) - (3.15), (3.20) - (3.23).

Proof. Let

$$\tau = t/T(\varepsilon) \quad \text{and} \quad \xi = \varepsilon e^{2\pi i\tau}\eta .$$

Then in the variables (τ,η) (3.24) becomes

$$2\pi i\eta + \frac{d\eta}{d\tau} = T(\varepsilon)\eta[\lambda(\mu) + \Sigma_1^{[\frac{L}{2}]} c_j(\mu)(\eta\bar{\eta})^j \varepsilon^{2j}] + O(\varepsilon^{L+1}) . \tag{3.28}$$

The assumed smoothness of $C(\xi,\bar{\xi},\mu)$ permits us to write the solution η, with $\eta(0) = 1$, in the form

$$\eta = \Sigma_0^L \eta_i \varepsilon^i + O(\varepsilon^{L+1}) , \text{ with } \eta_0(0) = 1 , \eta_i(0) = 0 \tag{3.29}$$

$$(1 \le i \le L) .$$

We shall show that

$$\eta_0 \equiv 1 , \quad \eta_i \equiv 0 \qquad (1 \le i \le L) .$$

We substitute the right-hand side of (3.29) for η in (3.28). Then at $O(\varepsilon^0)$ (3.28) yields

$$2\pi i\eta_0 + \frac{d\eta_0}{d\tau} = 2\pi i\eta_0 \quad \text{or} \quad \frac{d\eta_0}{d\tau} = 0 .$$

Hence

$$\eta_0 \equiv 1 .$$

At $O(\epsilon)$, (3.28) yields

$$2\pi i\eta_1 + \frac{d\eta_1}{d\tau} = 2\pi i\eta_1 + d_1 , \quad \text{or} \quad \frac{d\eta_1}{d\tau} = d_1 ,$$

where d_1 is a constant independent of ϵ . Thus

$$\eta_1 = d_1\tau + d_2 ,$$

where d_2 is a constant. But η , and hence η_1 , is 1-periodic. Consequently $d_1 = 0$; and since $\eta_1(0) = 0$, $d_2 = 0$. Therefore

$$\eta_1 \equiv 0 .$$

At $O(\epsilon^2)$, (3.28) yields

$$2\pi i\eta_2 + d\eta_2/d\tau = 2\pi i\eta_2 + d_3 \quad \text{or} \quad d\eta_2/d\tau = d_3 ,$$

where d_3 is a constant.. Then

$$\eta_2 = d_3\tau + d_4 .$$

But η_2 is 1-periodic since η is. Hence $d_3 = 0$; and since $\eta_2(0) = 0$, $d_4 = 0$. Therefore

$$\eta_2 \equiv 0 .$$

Continuing in this way, we show that

$$\eta_i = 0 \qquad (1 \le i \le L) .$$

If we now substitute the right-hand side of (3.25) for ξ in (3.24), then we obtain the already computed values (3.12) - (3.15) and (3.20) - (3.23) for $\mu_1,\ldots,\mu_4$ and $\tau_1,\ldots,\tau_4$ just as in the first part of this Section 3.

This completes the proof of the lemma.

Remark 1. The condition

$$d_1 = 0$$

that guarantees the $T(\varepsilon)$-periodicity of η_1 is the first of the orthogonality conditions that arise in Hopf's original paper [55], wherein the Poincaré-Lindstedt method is followed. We have not used this condition to evaluate μ_1 and τ_1 , nor have we used subsequent of the orthogonality conditions.

Remark 2. By the statement "$C(\xi,\bar{\xi},\mu)$ is C^{L+2} jointly in $\xi,\bar{\xi},\mu$" we mean that $C(\xi,\bar{\xi},\mu)$ is a function of the real variables $\xi_1 = \text{Re}\,\xi$, $\xi_2 = \text{Im}\,\xi$ and μ , and that the partial derivatives of C with respect to ξ_1, ξ_2 and μ of combined order $\le L + 2$ are all continuous. The variable $\bar{\xi}$ is included in order to indicate that the right-hand side of (3.24) is a function of ξ_1 and ξ_2 independently, not just in the combination $\xi_1 + i\xi_2$ as commonly understood in analytic function theory. $C(\xi,\eta,\mu)$ need not even be defined for $\eta \ne \bar{\xi}$. Our use of complex variables is merely for the computational convenience which the complex arithmetic provides.

4. STABILITY CRITERIA

In applications of the Hopf Bifurcation Theorem to systems modeling natural phenomena it is important to determine if the bifurcating periodic solutions are stable. We next present two approaches to the question of stability, one based on the Poincaré-Bendixson Theorem and the other a calculation of the Floquet exponent that determines stability. The approach using Floquet theory provides slightly more information.

We begin with the first mentioned approach. We assume that our 2×2 system has the Poincaré normal form (3.24) and that $\mu_2 \neq 0$, i.e., $\operatorname{Re} c_1(0) \neq 0$. Set $\xi = re^{i\theta}$. Then $\xi\bar{\xi} = r^2$ and

$$\frac{dr}{dt} = 2r \operatorname{Re} f(r,\theta,\mu) \ , \quad \frac{d\theta}{dt} = \operatorname{Im} f(r,\theta,\mu) \ , \tag{4.1}$$

where

$$f(r,\theta,\mu) = \lambda(\mu) + \Sigma_1^{[\frac{L}{2}]} c_j(\mu) r^{2j} + O(|(r,\mu)|^{L+1}) \ . \tag{4.2}$$

The periodic solution indexed by ε is just

$$r = \varepsilon + O(\varepsilon^{L+2}) \ , \quad \theta = 2\pi t/T(\varepsilon) + O(\varepsilon^{L+1}) \ .$$

Let $\delta_1 \in (0,1)$. Then there exists an $\varepsilon_2 < \varepsilon_P/\sqrt{2}$ such that for $0 < \varepsilon < \varepsilon_2$, the circle

$$C_1 = \{(r,\theta) \mid r^2 = \varepsilon^2(1 + \delta_1)\}$$

lies outside the orbit of the periodic solution indexed by ε ; and for any $\delta_2 \in (0,\frac{1}{2})$ the circle

$$C_2 = \{(r,\theta) \mid r^2 = \varepsilon^2\delta_2^2\}$$

lies inside the orbit of the periodic solution indexed by ε . There is no other periodic solution within the annulus determined by C_1 and C_2 ; see Fig. 1.4. Let P be any point

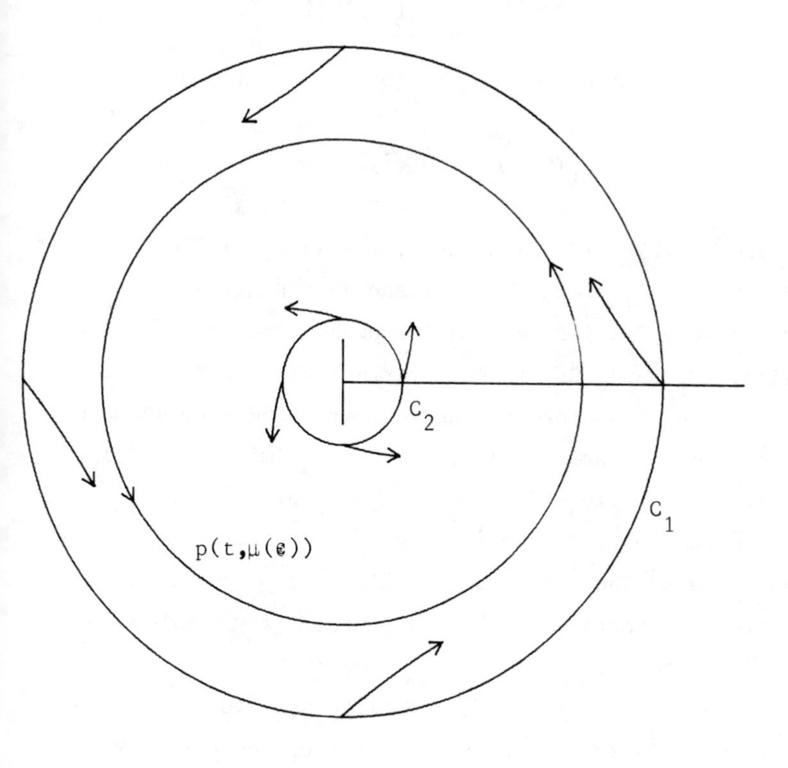

Figure 1.4.

n C_1 , and consider the trajectory of (4.1) passing through P .
n this trajectory at P

$$\frac{dr}{d\theta} = 2r \left.\frac{\mathrm{Re}\ f(r,\theta,\mu(\epsilon))}{\mathrm{Im}\ f(r,\theta,\mu(\epsilon))}\right|_{r=\epsilon(1+\delta_1)^{\frac{1}{2}}} .$$

For $r^2 = \epsilon^2(1+\delta_1)$,

$$\mathrm{Im}\ f(r,\theta,\mu(\epsilon)) = \omega(\mu(\epsilon)) + \Sigma_1^{[\frac{L}{2}]} \mathrm{Im}\ c_j(\mu(\epsilon))[\epsilon^2(1+\delta_1)]^j + O(\epsilon^{L+1})$$

$$= \omega_0 + O(\epsilon^2) ,$$

and

$$\operatorname{Re} f(r,\theta,\mu(\varepsilon)) = \operatorname{Re} \lambda(\mu(\varepsilon)) + \operatorname{Re} c_1(0)\varepsilon^2(1+\delta_1) + O(\varepsilon^4) ,$$

$$= \alpha'(0)\mu_2\varepsilon^2 + \operatorname{Re} c_1(0)\varepsilon^2(1+\delta_1) + O(\varepsilon^4) ,$$

$$= \operatorname{Re} c_1(0)\varepsilon^2\delta_1 + O(\varepsilon^4)$$

by (3.27) and (3.12) - (3.15). Hence, if $\operatorname{Re} c_1(0) < 0$, $dr/d\theta < 0$ at $P \in C_1$ for ε small enough. Similarly, $dr/d\theta > 0$ at $P \in C_2$ for ε small enough. Thus the annulus determined by C_1 and C_2 is an invariant set for (4.1) if $\operatorname{Re} c_1(0) < 0$, and therefore by the Poincaré-Bendixson Theorem [20], the periodic solution $p(t,\mu(\varepsilon))$ is a limit cycle (is asymptotically orbitally stable). If $\operatorname{Re} c_1(0) > 0$, then $p(t,\mu(\varepsilon))$ is instable, for we may apply the Poincaré-Bendixson Theorem as above to show that, in the limit $t \to -\infty$, $p(t,\mu(\varepsilon))$ is asymptotically orbitally stable; and hence, for increasing t, $p(t,\mu(\varepsilon))$ is instable.

Our second approach to stability is to use Floquet theory, which we now briefly review for the benefit of those readers unfamiliar with it; see G. Floquet [33], J. Hale [39, p. 118], or P. Hartman [43, p. 60]. We begin with a definition.

Definition. Let $p(t)$ be a T-periodic solution of $\dot{x} = f(x)$ ($p(t+T) = p(t)$ for all $t \in \mathbb{R}^1$), where $f \in C^1(\mathbb{R}^n,\mathbb{R}^n)$. Then p is asymptotically, orbitally stable with asymptotic phase if and only if there exists an $\varepsilon > 0$ such that if $\psi(t)$ is any solution of $\dot{x} = f(x)$ for which $|\psi(t_0) - p(t_0)| < \varepsilon$ at some time t_0, then there exists a constant ϕ, called the asymptotic phase, with the property

$$\lim_{t\to\infty} |\psi(t) - p(t+\phi)| = 0 .$$

Remark. If $x = \psi(t)$ satisfies $\dot{x} = f(x)$ and $y \equiv \psi(t) - p(t)$ is "small" then

$$\begin{aligned}\dot{y} &= f(p + y) - f(p) \\ &= A(t)y + o(|y|) ,\end{aligned}$$

where $A(t) = \partial f/\partial x$, evaluated at $p(t)$, is a square matrix with T-periodic elements. It turns out that the stability of p is largely determined by properties of solutions of the linear (variational) system obtained by neglecting the terms $o(|y|)$ in the above. This is our motivation for studying systems of the form (4.3) below.

Consider the linear, homogeneous, T-periodic system $(T > 0)$

$$\dot{y} = A(t)y \quad (A(t + T) = A(t) \quad \text{for all} \quad t \in \mathbb{R}^1) , \tag{4.3}$$

where A is a continuous real or complex $n \times n$ matrix function of t . Floquet's theory gives the structure of solutions of (4.3).

Definition. A *fundamental matrix solution* of (4.3) is a time-dependent matrix Y such that any solution vector y of (4.3) may be expressed as

$$y(t) = Y(t)c$$

for some constant vector c .

Theorem (Floquet). *Every fundamental matrix solution of (4.3) has the form*

$$Y(t) = P(t)e^{Bt} \qquad (-\infty < t < \infty) \tag{4.4}$$

for some $n \times n$ *T-periodic and differentiable matrix* $P(t)$ *and some constant matrix* B .

For a proof see Hale [39, p. 118].

Notes and Remarks

1. Suppose Y is a fundamental matrix solution of (4.3). Then $Y(t + T)$ is also such a solution, since A is T-periodic. Therefore, there exists a nonsingular matrix C such that

$$Y(t + T) = Y(t)C .$$

Hence there exists a matrix B such that

$$C = e^{BT} ;$$

see Hale [39, p. 118]. Now define

$$P(t) = Y(t)e^{-Bt} .$$

Then

$$\begin{aligned} P(t + T) &= Y(t + T)e^{-B(t+T)} \\ &= Y(t)e^{-Bt} \\ &= P(t) . \end{aligned}$$

Although C is uniquely determined by $A(t)$, B and $P(t)$ are not uniquely determined. For example, the representation (4.4) can be replaced by

$$Y(t) = [P(t)e^{-i\Omega t}]e^{(B+ i\Omega I)t} ,$$

where $\Omega = 2m\pi/T$ for some integer m.

2. The eigenvalues ρ_i of $C = e^{BT}$ are called the <u>characteristic multipliers</u> of (4.3). The eigenvalues β_i of B are called the <u>characteristic</u> (or <u>Floquet</u>) <u>exponents</u> of (4.3): $\rho_i = e^{\beta_i T}$. Their real parts are uniquely

determined by $A(t)$, while their imaginary parts are only determined $\bmod[2\pi/T]$.

3. If $\rho_j = \exp(\beta_j T)$ $(j = 1,\dots,n)$ are the characteristic multipliers of (4.3), then

$$\Pi_1^n \rho_j = \exp\left[\int_0^T \operatorname{tr} A(s)\, ds\right],$$

and

$$\Sigma_1^n \beta_j = \frac{1}{T}\int_0^T \operatorname{tr} A(s)\, ds \qquad (\text{mod } 2\pi i/T),$$

where

$$\operatorname{tr} A(s) = \Sigma_1^n A_{ii}(s).$$

4. One might conclude that linear, periodic systems are as simple in theory as linear systems with constant coefficients. However, the β_i are defined (known) only after the solutions of (4.3) are known; and there is, in general, no obvious relation or formula connecting the β_i with the coefficient matrix $A(t)$. Although it is not difficult numerically, determining the ρ_i (β_i) of linear, periodic systems <u>analytically</u> is an extremely difficult task except in the 2×2 case and for certain canonical systems and Hamiltonian systems; for amplification of this remark, see Hale [39].

5. Suppose one of the eigenvalues β of B is $0 \bmod 2\pi i/T$. Let v denote a corresponding eigenvector. Then

$$y(t) \equiv P(t)e^{Bt}v = P(t)e^{\beta t}v$$

is a T-periodic solution of (4.3). Conversely, if $y(t) \not\equiv 0$ is a T-periodic solution of (4.3), then, since we may write

$$y(t) = P(t)e^{Bt}c$$

for some constant vector c, the equation $y(t + T) = y(t)$ implies

$$e^{BT}c = c .$$

Thus c is an eigenvector of the matrix $C = \exp BT$ corresponding to an eigenvalue (characteristic multiplier) $\rho = 1$. By Remark 2 above, B therefore has an eigenvalue (characteristic exponent)

$$\beta = 0 \qquad \mod 2\pi i/T .$$

The following theorem connects Floquet Theory to the stability of periodic solutions of nonlinear, autonomous systems; see Hale [39, pp. 215-220] for the proof.

Theorem. *If* (1) $p(t)$ *is a nonconstant* T-*periodic solution of*

$$\dot{x} = f(x) \qquad (f \in C^1(\mathbb{R}^n,\mathbb{R}^n)) , \tag{4.5}$$

(2) *the characteristic multiplier* 1 *of the first variation of* (4.5) *with respect to the periodic solution* p, *namely of*

$$\frac{dy}{dt} = \frac{\partial f(p(t))}{\partial x} y , \tag{4.6}$$

is simple, (3) *all other characteristic multipliers of* (4.5) *have modulii less than* 1, *then* $p(t)$ *is asymptotically, orbitally stable with asymptotic phase*.

This theorem is H. Poincaré's in the analytic case.

We shall now apply this theorem to the real 2 by 2 system (3.1). We know from Section 3 that $\mu(\varepsilon)$ is C^{L+1} in

ϵ; hence $p_\epsilon(t) = \epsilon y(t,\epsilon,\mu(\epsilon))$ is C^{L+1} jointly in t and ϵ. Since $X = p_\epsilon(t)$ is a nonconstant, $T(\epsilon)$-periodic solution of (3.1), $\dot{p}_\epsilon(t)$ is a nontrivial, $T(\epsilon)$-periodic solution of the variational system $\dot{y} = A(t;\epsilon)y$, where $A(t;\epsilon) = \partial F/\partial X$ at $(p_\epsilon(t),\mu(\epsilon))$. By Remark 5 following Floquet's Theorem, one of the characteristic exponents associated with $p_\epsilon(t)$ is thus $0 \bmod 2\pi i/T(\epsilon)$. By Remark 3, $p_\epsilon(t)$ therefore has 0 and $\beta(\epsilon)$ as a set of characteristic exponents, where we <u>define</u>

$$\beta(\epsilon) = \frac{1}{T(\epsilon)} \int_0^{T(\epsilon)} \operatorname{tr} A(s;\epsilon)ds.$$

Since $T(\epsilon) = T(\epsilon,\mu(\epsilon))$ is C^{L+1} in ϵ and $A(t;\epsilon)$ is C^{L+1} jointly in t and ϵ, the function $\beta(\epsilon)$ is C^{L+1} in ϵ. Next we shall expand $\beta(\epsilon)$.

If we write $\xi = x_1 + ix_2$, then

$$\dot{x}_1 = \alpha x_1 - \omega x_2 + [(\text{Re } c_1)x_1 - (\text{Im } c_1)x_2]r^2 + O(\epsilon^4)$$

$$\dot{x}_2 = \omega x_1 + \alpha x_2 + [(\text{Re } c_1)x_2 + (\text{Im } c_1)x_1]r^2 + O(\epsilon^4)$$

and

$$\operatorname{tr} \frac{\partial F}{\partial X}(p(t,\mu(\epsilon)) = 2\alpha(\mu(\epsilon)) + 4[\text{Re } c_1(\mu(\epsilon))]\epsilon^2 + O(\epsilon^3),$$

where we have selectively used the facts that $\mu(\epsilon) = O(\epsilon^2)$ and $r^2 = \epsilon^2 + O(\epsilon^5)$, $(L \geq 2)$. Hence

$$\frac{1}{T(\epsilon)} \int_0^{T(\epsilon)} \operatorname{tr} A(s;\epsilon)ds = 2\alpha(\mu(\epsilon)) + 4[\text{Re } c_1(\mu(\epsilon))]\epsilon^2 + O(\epsilon^3).$$

But

$$\alpha(\mu(\epsilon)) = \alpha'(0)\mu_2\epsilon^2 + \dots = -\text{Re } c_1(0)\epsilon^2 + \dots .$$

so that by Remark 3 following Floquet's Theorem,

$$0 + \beta(\epsilon) = 2 \text{ Re } c_1(0)\epsilon^2 + O(\epsilon^3).$$

Thus $\beta(\epsilon) < 0$ for ϵ sufficiently small if $\text{Re } c_1(0) < 0$, which is just the criterion derived earlier for asymptotic, orbital stability of $p(t,\mu(\epsilon))$. However, the inequality $\beta(\epsilon) < 0$ implies that the bifurcating periodic solutions of the system (3.1) are asymptotically, orbitally stable _with asymptotic phase_.

The above computation can be carried out to include terms of order ϵ^4, provided L is at least 4. The result is

$$\beta(\epsilon) = \beta_2\epsilon^2 + \beta_4\epsilon^4 + O(\epsilon^5),$$

where

$$\beta_2 = 2 \text{ Re } c_1(0), \quad \beta_4 = 4 \text{ Re } c_2(0) + 2 \text{ Re } c_1'(0)\mu_2 .$$

Continuing in this fashion, we see that the expansion of $\beta(\epsilon)$ has the form (2.7c). This completes part I of our proof of the Hopf Bifurcation Theorem.

Remark. The reader should note that the Floquet exponent analysis above yields stability (instability) if $\beta(\epsilon) < 0$ ($\beta(\epsilon) > 0$), while the previous approach, using the Poincaré-Bendixson Theorem, only applies if $\beta_2 \neq 0$.

Remark. We call attention to Exercises 1-5 at the end of this Chapter; these illustrate Floquet theory.

5. REDUCTION OF TWO-DIMENSIONAL SYSTEMS TO POINCARÉ NORMAL FORM

Before we can apply the bifurcation formulae, derived in the preceding sections, to various model systems, we must show how general autonomous systems, satisfying the hypotheses of the Hopf Theorem, can be transformed into Poincaré normal form. We have subdivided this task into two parts. In this Section we begin with the single complex equation

$$\dot{z} = \lambda z + g(z,\bar{z};\mu) , \tag{5.1}$$

where

$$g(z,\bar{z};\mu) = \sum_{2 \le i + j \le L} g_{ij}(\mu) \frac{z^i \bar{z}^j}{i!j!} + O(|z|^{L+1}) \tag{5.2}$$

and

$$\lambda(\mu) = \alpha(\mu) + i\omega(\mu) .$$

Here $z = y_1 + iy_2$ is a complex variable; and (5.1) is equivalent to the real 2×2 system

$$\dot{y}_1 = f^1(y_1,y_2;\mu)$$

$$\dot{y}_2 = f^2(y_1,y_2;\mu) ,$$

with isolated stationary point at the origin and whose Jacobian matrix at the origin has the canonical form

$$D_y f(0,0,\mu) = \begin{pmatrix} \alpha(\mu) & -\omega(\mu) \\ \omega(\mu) & \alpha(\mu) \end{pmatrix} .$$

In Section 6 we shall show how to reduce general $n \times n$ systems to the form (5.1).

We desire to transform (5.1) by means of a transformation [46; 98]

$$z = \xi + \chi(\xi,\bar{\xi};\mu)$$

$$= \xi + \sum_{2\le i+j\le L} \chi_{ij}(\mu) \frac{\xi^i \bar{\xi}^j}{i!j!} , \qquad (5.3)$$

where $\chi_{ij} \equiv 0$ for $i = j + 1$, into the Poincaré normal form

$$\dot{\xi} = \lambda(\mu)\xi + \sum_{j=1}^{[\frac{L}{2}]} c_j(\mu)\xi|\xi|^{2j} + O(|\xi||(\xi,\mu)|^{L+1}) , \qquad (5.4)$$

$$= \lambda(\mu)\xi + \phi(\xi,\bar{\xi};\mu) .$$

We thank R. Ruppelt and A. Schneider for their careful analysis of this transformation [98,pp.35-43].

To use the bifurcation formulae for $\mu(\varepsilon)$, $T(\varepsilon)$, and $\beta(\varepsilon)$ derived previously we need only compute $c_1(0)$, $c_1'(0)$, and $c_2(0)$. In fact we shall compute $c_1(\mu)$ and $c_2(0)$.

We first proceed under the assumption that the transformation (5.3) taking (5.1) into (5.4) may be performed. After formally generating formulae for the coefficients in the expansion of ϕ , we shall prove that such a transformation is indeed possible.

By the chain rule,

$$\dot{z} = \dot{\xi} + \chi_\xi \dot{\xi} + \chi_{\bar{\xi}} \dot{\bar{\xi}} ,$$

or

$$\lambda\xi\chi_\xi + \bar{\lambda}\bar{\xi}\chi_{\bar{\xi}} - \lambda\chi = g(\xi + \chi, \bar{\xi} + \bar{\chi}) - (\phi + \chi_\xi\phi + \chi_{\bar{\xi}}\bar{\phi}) . \quad (5.5)$$

From this relation the coefficients χ_{ij} in (5.3) can be determined recursively. In powers of ξ and $\bar{\xi}$ the left-hand side of (5.5) can be written as

$$\sum_{2 \le i+j \le L} (\chi_{ij})(i\lambda + j\bar{\lambda} - \lambda) \frac{\xi^i \bar{\xi}^j}{i!j!} . \qquad (5.6)$$

Note that the expansion of the right-hand side of (5.5) to order $k = 2$ is independent of the χ_{ij} and to order k $(k = 3,\ldots,L)$ involves exactly the coefficients χ_{ij} for $2 \le i + j < k$. Therefore the undetermined coefficients χ_{ij} with $i + j = k$ can be found by expanding (5.5) to order k. We begin with $k = 2$.

To order $|\xi|^2$, (5.5) is

$$\lambda \chi_{20} \frac{\xi^2}{2} + \bar{\lambda} \chi_{11} \xi\bar{\xi} + (2\bar{\lambda} - \lambda)\chi_{02} \frac{\bar{\xi}^2}{2} = g_{20} \frac{\xi^2}{2} + g_{11}\xi\bar{\xi} + g_{02} \frac{\bar{\xi}^2}{2} .$$

Thus

$$\chi_{20} = \frac{g_{20}}{\lambda}, \quad \chi_{11} = \frac{g_{11}}{\bar{\lambda}}, \quad \chi_{02} = \frac{g_{02}}{2\bar{\lambda} - \lambda} . \qquad (5.7)$$

Taking this into account, we next find that equating coefficients of $\xi^2\bar{\xi}$ on both sides of (5.5) implies that

$$c_1(\mu) = \frac{g_{20}g_{11}(2\lambda + \bar{\lambda})}{2|\lambda|^2} + \frac{|g_{11}|^2}{\lambda} + \frac{|g_{02}|^2}{2(2\lambda - \bar{\lambda})} + \frac{g_{21}}{2} . \qquad (5.8)$$

Thus for $\mu = 0$ we obtain:

$$c_1(0) = \frac{i}{2\omega_0}\left(g_{20}g_{11} - 2|g_{11}|^2 - \tfrac{1}{3}|g_{02}|^2\right) + \frac{g_{21}}{2} , \qquad (5.9)$$

where we have used the hypothesis that $\lambda(0) = i\omega_0$.

We have now computed enough of χ so that if one is interested only in the values μ_2, τ_2, and β_2, then sufficient information is at hand. If one desires to compute μ_4, τ_4, and β_4, then $c_2(0)$ is needed, which requires the evaluation of χ_{ij} for $i + j \le 4$. Since we only desire to compute $c_2(0)$, we may set $\mu = 0$. Then $\lambda = i\omega_0$ and the $|\xi|^3$-terms of (5.5) imply that

$$\chi_{30} = \frac{-3i}{2\omega_0}(g_{20}\chi_{20} + g_{11}\bar{\chi}_{02} + \frac{g_{30}}{3}),$$

$$\chi_{12} = \frac{i}{2\omega_0}(g_{20}\chi_{02} + g_{11}(\bar{\chi}_{20} + 2\chi_{11}) + 2g_{02}\bar{\chi}_{11} + g_{12}), \quad (5.10)$$

$$\chi_{03} = \frac{3i}{4\omega_0}(g_{11}\chi_{02} + g_{02}\bar{\chi}_{20} + \frac{g_{03}}{3}),$$

where the χ_{ij} $(i + j = 2)$ are given by (5.7) with $\lambda = i\omega_0$. The next stage in the computation of $c_2(0)$, namely the computation of χ_{ij} for $i + j = 4$, is formidable. In the interest of accuracy we have done the computation of χ_{ij} for $i + j = 4$ both by hand and by means of a computer. Fortunately, we actually need only χ_{31}, χ_{22}, and χ_{13} of the χ_{ij} $(i + j = 4)$ to compute $c_2(0)$. The results are:

$$\chi_{31} = (6i/\omega_0)\, c_1(0)\chi_{20} - (i/\omega_0)(g_{31} + g_{11}\chi_{30} + g_{02}\bar{\chi}_{03})$$

$$- \frac{3i}{\omega_0}[g_{20}\chi_{11}\chi_{20} + g_{11}(\chi_{20}\bar{\chi}_{11} + \chi_{11}\bar{\chi}_{02} + \bar{\chi}_{12}) + g_{02}\bar{\chi}_{11}\bar{\chi}_{02} + g_{30}\chi_{11}$$

$$+ g_{21}(\chi_{20} + \bar{\chi}_{11}) + g_{12}\bar{\chi}_{02}],$$

$$\chi_{22} = (-8i/\omega_0)\,\mathrm{Re}\, c_1(0)\chi_{11} + \frac{i}{\omega_0}[g_{20}(\chi_{20}\chi_{02} + 2(\chi_{11})^2 + 2\chi_{12})$$

$$+ g_{11}(|\chi_{20}|^2 + 4|\chi_{11}|^2 + |\chi_{02}|^2) + g_{30}\chi_{02}$$

$$+ g_{02}(\bar{\chi}_{20}\bar{\chi}_{02} + 2(\bar{\chi}_{11})^2 + 2\bar{\chi}_{12}) + g_{21}(\bar{\chi}_{20} + 4\chi_{11}) + g_{12}\chi_{20} +$$

$$+ \; g_{03}\bar{\chi}_{02} + 4g_{12}\bar{\chi}_{11} + g_{22}] \,, \tag{5.11}$$

$$\chi_{13} = (-\frac{2i}{\omega_0})\bar{c}_1(0)\chi_{02} + (\frac{i}{3\omega_0})(g_{13} + g_{20}\chi_{03} + g_{11}\bar{\chi}_{30})$$

$$+ (\frac{i}{\omega_0})\,[g_{20}\chi_{11}\chi_{02} + g_{11}(\bar{\chi}_{20}\chi_{11} + \bar{\chi}_{11}\chi_{02} + \chi_{12}) + g_{02}\bar{\chi}_{20}\bar{\chi}_{11}$$

$$+ g_{21}\chi_{02} + g_{03}\bar{\chi}_{11} + g_{12}(\bar{\chi}_{20} + \chi_{11})] \,.$$

$$12c_2(0) = g_{20}(\chi_{30}\chi_{02} + 3\chi_{20}\chi_{12} + 3\chi_{22}) + g_{11}(\chi_{30}\bar{\chi}_{20} + \chi_{02}\bar{\chi}_{03} \tag{5.12}$$

$$+ 3\bar{\chi}_{22} + 2\chi_{31} + 6\chi_{11}\bar{\chi}_{12} + 3\chi_{12}\bar{\chi}_{02}) + g_{02}(\bar{\chi}_{20}\bar{\chi}_{03} + 6\bar{\chi}_{11}\bar{\chi}_{12}$$

$$+ 2\bar{\chi}_{13}) + 3g_{30}(\chi_{20}\chi_{02} + 2(\chi_{11})^2 + \chi_{12})$$

$$+ 3g_{03}(2\bar{\chi}_{11}\bar{\chi}_{02} + \bar{\chi}_{03}/3) + 3g_{21}(|\chi_{20}|^2 + 4|\chi_{11}|^2 + |\chi_{02}|^2$$

$$+ 2\chi_{20}\chi_{11}) + 3g_{12}(\bar{\chi}_{20}\bar{\chi}_{02} + 2\chi_{20}\bar{\chi}_{11} + 2\chi_{11}\bar{\chi}_{02} + 2(\bar{\chi}_{11})^2$$

$$+ 2\bar{\chi}_{12} + \chi_{30}/3) + g_{40}\chi_{02} + g_{31}(\bar{\chi}_{20} + 6\chi_{11}) + 3g_{22}(\chi_{20} + 2\bar{\chi}_{11})$$

$$+ 3g_{13}\bar{\chi}_{02} + g_{32} \,.$$

We now define

$$\chi(\xi,\bar{\xi};\mu) = \sum_{2 \le i+j+k \le L} \frac{\chi_{ij}^{(k)}(0)\xi^i\bar{\xi}^j\mu^k}{i!j!k!} \,, \tag{5.13}$$

where the coefficients $\chi_{ij}(\mu)$ are computed as described above. If the r.h.s. of (5.1) is C^{L+2} jointly in all arguments, then for all sufficiently small $|\mu|$ both g_{ij} and χ_{ij} are $C^{L-i-j+2}$ and hence c_i is C^{L-2i+1} in μ. Note that this function $\chi(\xi,\bar{\xi};\mu)$ is not <u>precisely</u> the same function used in (5.3), but this χ has better smoothness and produces the "same" normal form (5.4), i.e., the differences are hidden in the O-terms.

The argument just presented is formal because existence, smoothness, and the form of the expansion of ϕ in (5.4) were assumed <u>a priori</u>. We now prove $\phi(\xi,\bar{\xi};\mu)$ has these properties.

Returning to (5.5) with $\chi(\xi,\bar{\xi};\mu)$ now known, we observe that (5.5) may be rewritten in the form

$$F(\mathrm{Re}\phi,\ \mathrm{Im}\phi,\ \mathrm{Re}\xi,\ \mathrm{Im}\xi;\mu) = 0\ ,$$

where F is at least C^1 in all its arguments and $F(0,0,0,0;0) = 0$. Moreover,

$$\begin{pmatrix} \dfrac{\partial \mathrm{Re}\ F}{\partial \mathrm{Re}\ \phi} & \dfrac{\partial \mathrm{Re}\ F}{\partial \mathrm{Im}\ \phi} \\ \\ \dfrac{\partial \mathrm{Im}\ F}{\partial \mathrm{Re}\ \phi} & \dfrac{\partial \mathrm{Im}\ F}{\partial \mathrm{Im}\ \phi} \end{pmatrix}$$

at $(0,0,0,0;0)$ is

$$\begin{pmatrix} 1 & 0 \\ 0 & 1 \end{pmatrix}.$$

Thus, by the Implicit Function Theorem, there exists a unique $\phi = \phi(\xi,\bar{\xi};\mu)$, $\phi\colon \mathbb{C}\times\mathbb{C}\times I \to \mathbb{C}$, defined and C^1 in all arguments in a neighborhood of $(0,0;0)$, and such that $\phi(0,0;0) = 0$ and (5.5) holds. Since for fixed μ both g and χ are C^{L+2} in the remaining variables, so is ϕ. Thus ϕ has a Taylor expansion of the form

$$\phi(\xi,\bar{\xi};\mu) = \sum_{2\le i+j\le L+1} \frac{\phi_{ij}(\mu)}{i!j!} \xi^i \bar{\xi}^j + O(|\xi|^{L+2}) ,$$

and the bound implied for the O-term is uniform in μ for all sufficiently small $|\mu|$. Further, by direct computation, we find that

$$\phi_{ij}(\mu) = O(|\mu|^{L-i-j+2}) \qquad (i \ne j+1,\ 2 \le i+j \le L+1)$$

and

$$\phi_{j+1,j}(\mu) = (j+1)!j!\, c_j(\mu) + O(|\mu|^{L-2j+1}) \qquad (1 \le j \le [\tfrac{L}{2}]) .$$

Thus the equation (5.1) is transformed by means of (5.3), with the function χ defined as above, into an equation of the form (5.4). Since the transformation χ is a polynomial in each of its arguments, the right-hand side of the normal form (5.4) that results is C^{L+2} jointly in $\xi, \bar{\xi}$, and μ. This completes Part II of our proof of the Hopf Bifurcation Theorem (Theorem II of Section 2).

It is a natural question, whether the assumption that $g(z,\bar{z};\mu)$ is analytic implies convergence of the infinite series

$$\sum_{2\le i+j} \chi_{ij}(\mu) \frac{\xi^i \bar{\xi}^j}{i!j!} , \quad \xi \sum_{i=1}^{\infty} c_i(\mu)(\xi\bar{\xi})^i .$$

For a discussion of questions of this kind see Arnol'd [6, Chapters 5 and 6].

6. RESTRICTION TO THE CENTER MANIFOLD

In this Section we complete our proof of the Hopf Bifurcation Theorem by showing how to reduce the n-dimensional system (2.1) to the two-dimensional one in the standard form (5.1). The major step in this reduction is an application of the Center Manifold Theorem [72, 81, 97] to the suspended system

$$\begin{aligned} \dot{X} &= F(X;\mu) \\ \dot{\mu} &= 0 \end{aligned} \tag{6.1}$$

at $(X,\mu) = (0,0) \in \mathbb{R}^n \times \mathbb{R}^1$. A center manifold $\mathcal{C}$ for this system is a locally invariant, locally attracting, three-dimensional manifold in $\mathbb{R}^n \times \mathbb{R}^1$, containing the origin and an interval $\{|\mu| < \mu_0\}$ of the μ-axis (we assume $F(0,\mu) = 0$ for $|\mu| < \mu_0$) and tangent at the origin to the three-dimensional subspace spanned by the eigenvectors of the matrix

$$\begin{pmatrix} F_X(0;0) & 0 \\ 0 & 0 \end{pmatrix}$$

corresponding to the eigenvalues 0, $i\omega_0$ and $-i\omega_0$.

Local invariance means that a trajectory of a solution with initial point on $\mathcal{C}$ remains on $\mathcal{C}$ (at least for a small time) and local attraction means that there is a neighborhood of the origin with the property, any trajectory which remains within this neighborhood as $t \to \infty$ necessarily approaches $\mathcal{C}$.

The Center Manifold Theorem we apply is P. Hartman's version [43, Chapter IX] of A. Kelley's original theorem [72]. Hartman's version applies to autonomous systems of ordinary differential equations and guarantees the existence of C^K center manifolds under C^K hypotheses on F. The Center Manifold Theorem we prove in Appendix A applies to delay-differential and partial differential equations, as well as to ordinary differential equations. When applied to (6.1), this more general theorem, however, only establishes existence of C^{K-1} center manifolds under C^K hypotheses on F.

A second difference between Hartman's Center Manifold Theorem and the more general one proved in Appendix A is that to apply Hartman's theorem we must transform (6.1) into a canonical form. These differences create a minor dilemma: on the one hand, to obtain the C^{L+2}-smoothness of the center manifold(s) that we need to prove Theorem II, we must transform (6.1) to canonical form; but, on the other hand, we want to avoid reduction to this canonical form so that the restriction of (6.1) to the center manifold(s) carried out here will be a model for analysis of Hopf bifurcation for delay-differential and partial differential equations in Chapters 4 and 5. To release ourselves from this dilemma, we first apply Hartman's theorem to (6.1), and then we back up and proceed along a different path, the one to be followed in later chapters.

Note. Although center manifolds are not necessarily unique, we regard this potential nonuniqueness as a mathematical fine-point. Each center manifold contains all local recurrent behaviour of the system. Since bifurcating periodic solutions are local recurrent behaviour, we may obtain this family of solutions within any representative center manifold. Computationally there is no difficulty, for it is Taylor expansions we shall use; and Wan has shown [108] that the Taylor expansions are the same for each center manifold. Hereafter when we refer to the center manifold, it is to be understood that we are selecting a member of the equivalence class of center manifolds to represent the class.

Remark. For the explanation of why the Center Manifold Theorem is applied to the suspended system (6.1) rather than to the original system (2.1) at $\mu = 0$, see Exercises 8 and 9 at the end of this Chapter.

Let $q(\mu)$ and $q^*(\mu)$ be eigenvectors for

$$A(\mu) = F_X(0,\mu)$$

and A^T, respectively, corresponding to the simple eigenvalues

$$\lambda(\mu) = \alpha(\mu) + i\omega(\mu) \quad \text{and} \quad \overline{\lambda}(\mu)$$

of $A(\mu)$; that is

$$Aq = \lambda q \ , \quad A^T q^* = \overline{\lambda} q^* \ . \tag{6.2}$$

We normalize q^* relative to q so that

$$\langle q^*, q \rangle = 1 \ ,$$

where $\langle \cdot \, , \cdot \rangle$ denotes the Hermitian product

$$\langle u,v \rangle = \Sigma_1^n \, \overline{u}_i v_i \ .$$

To put the system (6.1) into the cononical form required by the versions [43, 72] of the Center Manifold Theorem, we let P_0 denote any real $n \times n$ matrix whose first column is $\mathrm{Re}\, q(0)$, whose second column is $-\mathrm{Im}\, q(0)$, and whose last $n - 2$ columns are any set of real, linearly independent vectors e_j, $(1 \le j \le n - 2)$ obeying the condition $\langle q^*(0), e_j \rangle = 0$, $(1 \le j \le n - 2)$. Under the change of variables $X = P_0 Y$ the system (6.1) assumes the form

$$\dot{Y} = \begin{pmatrix} 0 & -\omega_0 & 0 \\ \omega_0 & 0 & \\ 0 & & D_0 \end{pmatrix} Y + G(Y;\mu)$$

$$\dot{\mu} = 0 \ ,$$

where $G(Y;\mu) = O(|(Y,\mu)|^2)$ and is C^{L+2} jointly in Y, μ, and D_0 is real, $(n - 2) \times (n - 2)$, and has as eigenvalues $\lambda_3(0), \ldots, \lambda_n(0)$. The vectors e_j, $(1 \le j \le n - 2)$ can be chosen so that D_0 is in real canonical form as required by [72]

We can picture a center manifold for this system as in Fig. .6(a) below. Near the origin in y_1, y_2, μ-space, if $\mu_2 \neq 0$, he bifurcating periodic solutions themselves form (roughly) a arabolic surface; see Fig. 1.6(b).

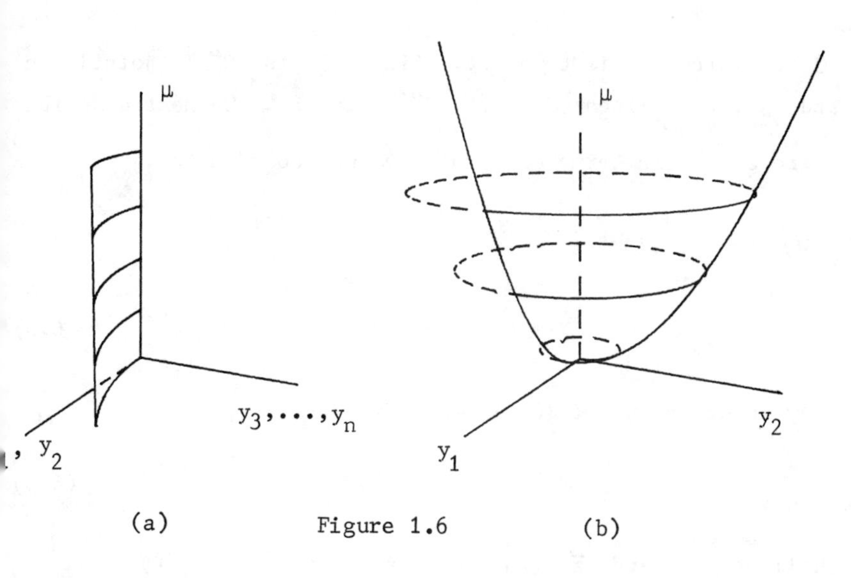

(a) Figure 1.6 (b)

For this canonical system there exists a center manifold

$$C = \{(Y,\mu) \mid (Y,\mu) = (y_1, y_2, W(y_1, y_2, \mu), \mu), \ |(y_1, y_2, \mu)| < \delta\},$$

or some sufficiently small δ, where

$$W: \mathbb{R}^3 \to \mathbb{R}^{n-2}$$

s C^{L+2} jointly in y_1, y_2, and μ and $W(y_1, y_2, \mu) =$ $(|(y_1, y_2, \mu)|^2)$. In terms of the X-coordinates

$$C = \{(X,\mu) \mid (X,\mu) = (y_1 \mathrm{Re}\, q(0) - y_2 \mathrm{Im}\, q(0)$$

$$+ \sum_{j=1}^{n-2} e_j W_j, \mu), \ |(y_1, y_2, \mu)| < \delta\}.$$

However, to determine the restriction of the system (6.1) to $\mathcal{C}$, we find it more convenient to describe $\mathcal{C}$ in terms of slices $\mathcal{C}_\mu$,

$$\mathcal{C}_\mu = \{X|\ (X,\mu) \in \mathcal{C}\} ,$$

of $\mathcal{C}$ on which μ is constant. Since $\mathcal{C}$ is C^{L+2} jointly in X and μ, the slices $\mathcal{C}_\mu$ are C^{L+2} in X. We next describe the slices $\mathcal{C}_\mu$ in terms of complex-valued coordinates.

Writing (2.1) as

$$\dot{X} = A(\mu)X + f(X,\mu) , \tag{6.3}$$

for any solution x of (6.3), we define

$$z(t) = \langle q^*, x(t)\rangle . \tag{6.4}$$

We shall use z and $\bar{z}$ (in the directions q and $\bar{q}$) as local coordinates. We also define

$$\begin{aligned} w(t) &= x(t) - z(t)q(\mu) - \bar{z}(t)\bar{q}(\mu) \\ &= x(t) - 2\mathrm{Re}[z(t)q(\mu)] . \end{aligned} \tag{6.5}$$

Because zq and $\bar{z}\bar{q}$ will always occur together, our choice of complex coordinates will not introduce complex-valued solutions of (6.3).

<u>Remark</u>. In order to calculate μ_2, τ_2 and β_2 relative to (2.1), only $\lambda(0)$, $q(0)$, and $q^*(0)$ are actually required.

<u>Remark</u>. Note that $\langle q^*,\bar{q}\rangle = 0$. For example, if at $\mu = 0$

$$A = \begin{bmatrix} 3 & 2 & -3 \\ -4 & 0 & 2 \\ 6 & 2 & -6 \end{bmatrix},$$

then

$$\lambda(0) = i\omega_0 = 2i \ , \quad q(0) = \begin{pmatrix} 1 \\ i \\ 1 \end{pmatrix}, \quad \text{and} \quad q^*(0) = \begin{pmatrix} 1 \\ i/2 \\ -1/2 \end{pmatrix} .$$

(The real eigenvalue is $\lambda_3 = -3$) .

Remark. A slightly different way of looking at the decomposition of $x(t)$ into $z(t)$ and $w(t)$ is by means of the projection matrices

$$P_{\parallel} = q(\overline{q}^*)^T + \overline{q}(q^*)^T ,$$

$$= 2 \operatorname{Re}[q(\overline{q}^*)^T]$$

and

$$P_{\perp} = I - P_{\parallel} ,$$

$$= I - 2\operatorname{Re}[q(\overline{q}^*)^T] .$$

These obey

$$P_{\parallel}^2 = P_{\parallel} , \quad P_{\perp}^2 = P_{\perp} , \quad P_{\parallel} P_{\perp} = 0 , \quad P_{\perp} P_{\parallel} = 0 .$$

In terms of $P_{\parallel}$ and $P_{\perp}$

$$z(t)q + \overline{z}(t)\overline{q} = P_{\parallel} x(t)$$

and

$$w(t) = P_{\perp} x(t) .$$

In the variables z and w, (6.3) becomes

$$\begin{aligned} \dot{z} &= \lambda(\mu) z + G(z,\bar{z},w;\mu) \\ \dot{w} &= A(\mu) w + H(z,\bar{z},w;\mu) , \end{aligned} \tag{6.6}$$

where

$$\begin{aligned} G(z,\bar{z},w;\mu) &= \langle q^*, f(w + 2\mathrm{Re}[zq];\mu) \rangle \\ H(z,\bar{z},w;\mu) &= f(w + 2\mathrm{Re}[zq];\mu) - 2\mathrm{Re}[qG] . \end{aligned} \tag{6.7}$$

Since

$$\begin{aligned} &\langle \mathrm{Re}\, q^*, w \rangle = 0 \quad \text{and} \quad \langle \mathrm{Im}\, q^*, w \rangle = 0 , \\ &\langle q^*, w \rangle = 0 . \end{aligned} \tag{6.8}$$

Although (6.6) appears to have (real) dimension $n + 2$, the above orthogonality relations imply that two components of w are linear combinations of the other components (with time independent coefficients). We have adopted the form (6.6) here because it readily generalizes to differential-difference and partial differential equations, which we shall treat in Chapters 4 and 5.

Since $F(X;\mu) \in C^{L+2}$, $q(\mu)$ and $q^*(\mu)$ are C^{L+1} in μ, and $f(x,\mu)$ is C^{L+1} jointly in all its arguments. Therefore, the family of manifolds $\mathcal{C}_{\mu}$ $(|\mu| < \delta)$ may be locally represented as a real vector-valued function

$$w = w(z,\bar{z};\mu) ,$$

that is C^{L+1} jointly in all its arguments and satisfies

$$w_z(0,0;\mu) = w_{\bar{z}}(0,0;\mu) = 0 \ , \quad \text{and} \quad \langle q^*,w\rangle = 0 \ .$$

(By "$f(z,\bar{z})$ is C^k jointly in z and $\bar{z}$" we mean only that the quantities $(\partial/\partial z)^i(\partial/\partial\bar{z})^j f(y_1 + iy_2, y_1 - iy_2)$ $(0 \le i + j \le k)$ are continuous in y_1 and y_2, where the operators $\partial/\partial z$ and $\partial/\partial\bar{z}$ are defined by

$$\frac{\partial}{\partial z} = \frac{1}{2}(\frac{\partial}{\partial y} - i\frac{\partial}{\partial y_2}) \ , \ \frac{\partial}{\partial \bar{z}} = \frac{1}{2}(\frac{\partial}{\partial y_1} + i\frac{\partial}{\partial y_2}) \ .)$$

Thus for $|\mu| < \delta$ the intersection of $\mathcal{C}$, restricted to R^n with some neighborhood of the origin has the form

$$\mathcal{C}_\mu = \{x \in R^n \mid x = w(z,\bar{z};\mu) + 2\,\mathrm{Re}[zq(\mu)] \ , \quad |z| < \delta\} \ .$$

Since the periodic solutions we wish to describe represent recurrent behavior for (6.6) and all such behavior is included in the center manifold $\mathcal{C}$ [72], we now restrict (6.6) to $\mathcal{C}_\mu$ by setting

$$w(t) = w(z(t),\bar{z}(t);\mu) \ . \tag{6.9}$$

The system on $\mathcal{C}_\mu$ is described by

$$\dot{z} = \lambda z + g(z,\bar{z};\mu) \ , \tag{6.10}$$

where

$$g(z,\bar{z};\mu) = G(z,\bar{z},w(z,\bar{z};\mu);\mu) \tag{6.11}$$

and $w(z,\bar{z};\mu)$ is determined by

$$\dot{w} = w_z(\lambda z + g) + w_{\bar{z}}(\bar{\lambda}\bar{z} + \bar{g}) = Aw + H(z,\bar{z},w;\mu) . \qquad (6.12)$$

The right-hand side of (6.10) is C^{L+1} jointly in z, $\bar{z}$, and μ; and the function $g(z,\bar{z};\mu)$ is C^{L+1} jointly in z and $\bar{z}$ and, for fixed μ, satisfies

$$g_z(0,0;\mu) = g_{\bar{z}}(0,0;\mu) = 0 ,$$

Hence (6.10) is of the desired form (5.1).

The careful reader will note that there was a loss of smoothness in the process of defining $w(z,\bar{z},\mu)$, a loss due to our choice of variables z, $\bar{z}$, where

$$z = \langle q^*,x \rangle ,$$

since $q^*(\mu)$ is only C^{L+1} in μ. One may avoid this loss of smoothness by using the expansion

$$\sum_{j=0}^{L+1} q^{*(j)}(0)\mu^j/j!$$

in place of $q^*(\mu)$. If one does this, the right-hand side of (6.10) is C^{L+2} jointly in z, $\bar{z}$, and μ, as claimed; and the partial derivatives of g of combined order not greater than $L + 1$ at $z = \bar{z} = \mu = 0$ are unchanged.

To complete our proof of Theorem II we now show how to relate the stability of the bifurcating periodic solutions within $\mathbb{R}^n$ to the stability (shown in Section 4) of these solutions within $\mathbb{C}_\mu$. The characteristic exponent $\beta(\varepsilon)$ in Section 4 was obtained from a two-dimensional system of ordinary differential equations. In fact, $\beta(\varepsilon)$ is also a characteristic exponent for the n-dimensional system (6.1): the verification of this claim follows.

Let x_0 denote a point on the orbit of the n-dimensional

solution $p_\varepsilon(t)$, say $p_\varepsilon(t_0)$. Let e_n denote the unit vector in the direction of $\dot{p}_\varepsilon(t_0) = F(x_0;\mu(\varepsilon))$, and let Π_0 denote the hyperplane

$$\Pi_0 = \{x \mid (x - x_0)\cdot e_n = 0\} .$$

The hyperplane Π_0 may be also be defined in terms of a set of orthogonal unit vectors $\{e_1,e_2,\ldots,e_{n-1}\}$, with the property

$$e_j\cdot e_n = 0 ; \quad (j = 1,2,\ldots,n - 1) ;$$

$$\Pi_0 = \{x = x_0 + \Sigma_{j=1}^{n-1}\eta_j e_j , \quad \eta \in \mathbb{R}^{n-1}\} .$$

Consider the mapping defined as follows: integrate the system

$$\dot{x} = F(x; \mu(\varepsilon))$$

$$x(0) = x_0 + \Sigma_{j=1}^{n-1}\eta_j e_j$$

until the time $t = \tau(\eta)$ nearest to $T(\varepsilon)$ at which the solution obeys

$$x(\tau(\eta)) \in \Pi_0 .$$

Then let $\psi = \psi(\eta)$ be the vector such that

$$x(\tau(\eta)) = x_0 + \Sigma_{j=1}^{n-1}\psi_j(\eta)e_j .$$

It is shown in [43], for example, that there is a neighborhood $\mathfrak{n}$ of 0 in $\mathbb{R}^{n-1}$ such that for all η in this neighborhood, the map $\psi: \eta \to \mathbb{R}^{n-1}$, called the Poincaré map, is well defined. Moreover [43; pp. 251-253], it is shown that the eigenvalues of the matrix $\partial\psi/\partial\eta(0)$ are precisely $n - 1$ of the characteristic multipliers associated with the periodic solution (the remaining multiplier is identically 1 and is associated with the solution $\dot{p}_\varepsilon(t)$ of the variational equation).

Now $\mathbb{C}_\mu$ is a two-dimensional invariant manifold for the system $\dot{x} = F(x;\mu)$, and the orbit of $p_\varepsilon(t)$ lies in $\mathbb{C}_\mu$. The manifold $\mathbb{C}_\mu$ itself intersects the hyperplane Π_0, and the intersection describes a curve Γ in $\mathbb{R}^n$. Since Γ belongs to Π_0, it may be parametrized as

$$\Gamma = \{x_s = x_0 + \sum_{j=1}^{n-1} \eta_j(s) e_j \ , \quad |s| < \sigma\}$$

or as

$$\Gamma^{n-1} = \{\eta = \eta(s) = (\eta_1(s), \eta_2(s), \ldots, \eta_{n-1}(s))^T \ , \quad |s| < \sigma\} ,$$

where $\eta_j(0) = 0$ $(j = 1,\ldots,n-1)$, $\sigma > 0$ is sufficiently small, and s denotes arclength. Corresponding to each $x_s \in \mathbb{C}$, there are points $z_s \in \mathbb{C}$ and $\xi_s \in \mathbb{C}$ such that

$$x_s = z_s q + \bar{z}_s \bar{q} + w(z_s, \bar{z}_s; \mu) ,$$

$$z_s = \xi_s + \chi(\xi_s, \bar{\xi}_s; \mu) .$$

Consider the tangent vector

$$v = \frac{d}{ds} (\eta(s)) \Big|_{s=0} .$$

This vector is, we claim, an eigenvector of the Jacobian matrix $\partial\psi/\partial\eta(0)$, with eigenvalue $\exp(\beta(\varepsilon)T(\varepsilon))$. To see this we let $\tilde{\eta}(s) = \psi(\eta(s))$, and let $\tilde{x}_s$, $\tilde{z}_s$ and $\tilde{\xi}_s$ correspond to $\tilde{\eta}(s)$. The definition of $\beta(\varepsilon)$ implies that

$$\lim_{s \to 0} \frac{|\tilde{\xi}_s - \xi_0|}{|\xi_s - \xi_0|} = \exp(\beta(\varepsilon)\ T(\varepsilon)) ,$$

and a straightforward computation using Taylor expansions shows that

$$\lim_{s\to 0} \frac{|\tilde{\eta}(s)|}{|\eta(s)|} = \lim_{s\to 0} \frac{|\tilde{z}_s - z_0|}{|z_s - z_0|} = \lim_{s\to 0} \frac{|\tilde{\xi}_s - \xi_0|}{|\xi_s - \xi_0|} = \exp(\beta(\varepsilon)T(\varepsilon)).$$

Since $\tilde{\eta}(s) = \psi(\eta)(s)) \in \Gamma^{n-1}$, we may write $\tilde{\eta}(s) = \eta(\tilde{s}(s))$ for some $\tilde{s} = \tilde{s}(s)$. By the geometry of Γ,

$$\lim_{s\to 0} \frac{\tilde{s}(s)}{s} = \lim_{s\to 0} \frac{|\tilde{\eta}(s)|}{|\eta(s)|} = \exp(\beta(\varepsilon)T(\varepsilon)) .$$

Now, by Taylor expansion of the map ψ,

$$\tilde{\eta}(s) = \psi(\eta(s)) = \frac{\partial\psi}{\partial\eta}(0)\, \eta(s) + O(|s|^2) ,$$

$$\frac{\tilde{s}}{s}\frac{\eta(\tilde{s})}{\tilde{s}} = \frac{\partial\psi}{\partial\eta}(0)\frac{\eta(s)}{s} + O(|s|) ;$$

and therefore, in the limit $s \to 0$,

$$\exp(\beta(\varepsilon)T(\varepsilon))\, v = \frac{\partial\psi}{\partial\eta}(0)\, v$$

as claimed. Note that this result provides a geometric interpretation of the characteristic exponent $\beta(\varepsilon)$: the unit eigenvector of $\partial\psi/\partial\eta(0)$ corresponding to the eigenvalue $\exp(\beta(\varepsilon)T(\varepsilon))$ is the unit tangent v to the curve Γ^{n-1} at the origin.

There are precisely n characteristic multipliers associated with the periodic solution $p_\varepsilon(t)$: one of these is identically 1, and is associated with the solution $\dot{p}_\varepsilon(t)$ of the variational system. As we have argued above, another is $\exp(\beta(\varepsilon)T(\varepsilon))$. The remaining $n - 2$ multipliers have moduli which are strictly less than 1 for small ε, for

$$\rho_j = \exp(2\pi\lambda_j(0)/\omega(0)) + o(1) , \quad (j = 3,\ldots,n) ;$$

and, by the Hopf hypotheses, the eigenvalues $\lambda_j(0)$ $(j = 3,\ldots,n)$ have strictly negative real parts.

Our proof of the Hopf Bifurcation Theorem (Theorem II) is now complete. There remains, however, the practical task of computing the coefficients in the expansions (6.9) and

$$g(z,\bar{z};\mu) = \sum_{i+j=2}^{L+1} \frac{g_{ij}(\mu)}{i!j!} z^i \bar{z}^j + O(|z|^{L+2}) . \tag{6.13}$$

To accomplish this we rewrite (6.12) as

$$Lw = H(z,\bar{z},w;\mu) - g w_z - \bar{g} w_{\bar{z}} , \tag{6.14}$$

$$= H(z,\bar{z},w(z,\bar{z};\mu);\mu) - 2\mathrm{Re}[G(z,\bar{z},w(z,\bar{z};\mu);\mu) w_z(z,\bar{z};\mu)] ,$$

where

$$L \equiv (\lambda z \frac{\partial}{\partial z} + \bar{\lambda}\bar{z} \cdot \frac{\partial}{\partial \bar{z}} - A) .$$

Now

$$Lw = \sum_{i+j=2}^{L+1} [(\lambda i + \bar{\lambda} j)I - A] \frac{w_{ij}(\mu)}{i!j!} z^i \bar{z}^j + O(|z|^{L+2}) , \tag{6.15}$$

and we can write the right-hand side of (6.14) as

$$\sum_{i+j=2}^{L+1} r_{ij} \frac{z^i \bar{z}^j}{i!j!} + O(|z|^{L+2}) .$$

To solve the equations

$$[(\lambda i + \bar{\lambda} j)I - A] w_{ij}(\mu) = r_{ij}(\mu) \tag{6.16}$$

for the w_{ij} in terms of the r_{ij} we must investigate the coefficient matrices

$$B_{ij} \equiv [(\lambda i + \bar{\lambda} j)I - A] .$$

For fixed i,j $(2 \leq i + j \leq L + 1)$, the eigenvalues of B_{ij} are

$$\lambda i + \bar{\lambda} j - \lambda_k \qquad (k = 1,\ldots,n) .$$

For all sufficiently small $|\mu|$, these eigenvalues have strictly positive real parts, except possibly for $k = 1,2$ $(\lambda_1 = \lambda, \lambda_2 = \bar{\lambda})$, by our hypotheses in the Hopf Bifurcation Theorem (see Section 2). The eigenvectors of B_{ij} for $k = 1,2$ are q and $\bar{q}$, respectively; and the right-hand side of (6.14) obeys the orthogonality conditions

$$\langle \mathrm{Re}\, q^* , \quad H(z,\bar{z},w;\mu) - 2\, \mathrm{Re}(gw_z) \rangle = 0$$

and

$$\langle \mathrm{Im}\, q^* , \quad H(z,\bar{z},w;\mu) - 2\, \mathrm{Re}(gw_z) \rangle = 0 .$$

Thus the coefficient r_{ij} in the expansion of the right-hand side of (6.16) also satisfies these orthogonality conditions. Hence there exists a unique solution $w_{ij}(\mu)$ of (6.16) such that

$$\langle \mathrm{Re}\, q^*, w_{ij} \rangle = \langle \mathrm{Im}\, q^*, w_{ij} \rangle = 0 ,$$

even in those cases where B_{ij} is not invertible. Now B_{ij} is not invertible if either

$$\lambda_i + \bar{\lambda} j - \lambda_1 = \lambda(i - 1) + \bar{\lambda} j = 0$$

or

$$\lambda i + \bar{\lambda} j - \lambda_2 = \lambda i + \bar{\lambda}(j - 1) = 0 .$$

This happens for $\mu = 0$ and $i = j \pm 1$, i.e., only for $\mu = 0$

and $i + j = 3,5,\ldots$.

For sufficiently smooth functions $\phi(z,\bar{z})$ and $p \geq 0$ we define $\mathfrak{J}_p\phi$ be the homogeneous polynomial of degree p in the Taylor series expansion of ϕ about $(0,0)$. We further define

$$g_p(z,\bar{z};\mu) = \mathfrak{J}_p g \; , \quad w_p(z,\bar{z};\mu) = \mathfrak{J}_p w$$

$$(0 \leq p \leq L + 1) \; .$$

Then the algorithm for determining the expansion coefficients g_{ij} and w_{ij} $(2 \leq i + j \leq L + 1)$ in (6.9) and (6.13) is described by

$$g_0 = g_1 = w_0 = w_1 = 0 \; ,$$

and

$$g_p(z,\bar{z};\mu) = \mathfrak{J}_p G(z,\bar{z},w^{p-1}(z,\bar{z};\mu);\mu) \; ,$$

where $w^{p-1}(z,\bar{z};\mu) = \Sigma_z^{p-1} w_k(z,\bar{z};\mu)$, while w_p is the unique solution of

$$\begin{cases} L\, w_p = \mathfrak{J}_p[H(z,\bar{z},w^{p-1};\mu) - g^{p-1}\dfrac{\partial w^{p-1}}{\partial z} - \bar{g}^{p-1}\dfrac{\partial w^{p-1}}{\partial \bar{z}} \\ \langle q^*, w_p \rangle = 0 \end{cases}$$

in which $g^{p-1}(z,\bar{z};\mu) = \Sigma_2^{p-1} g_k(z,\bar{z};\mu)$. For $p = 2$, the algorithm produces $g_2 = \mathfrak{J}_2 G(z,\bar{z},0;\mu)$, or

$$g_{ij} = G_{ij}^0 = \frac{\partial^2}{\partial z^i \partial \bar{z}^j} G(0,0,0;\mu) \qquad (i + j = 2) \; ,$$

and

$$w_{ij} = [(\lambda i + \bar{\lambda} j)I - A]^{-1} H_{ij}^0 \qquad (i + j = 2) \; .$$

where

$$H_{ij}^{0} \equiv \frac{\partial^2}{\partial z^i \partial \bar{z}^j} H(0,0,0;\mu) .$$

For $p = 3$, the algorithm yields

$$g_3 = \mathfrak{J}_3 G(z,\bar{z},w_2(z,\bar{z};\mu);\mu) .$$

In particular,

$$g_{21} = G_{21}^{0} + G_{01}^{1} w_{20} + 2G_{10}^{1} w_{11} ,$$

where

$$G_{01}^{1} = \frac{\partial}{\partial \bar{z}} \left(\frac{\partial G}{\partial w^1}, \ldots, \frac{\partial G}{\partial w^n}\right) \Bigg|_{(0,0,0;\mu)}$$

and

$$G_{10}^{1} = \frac{\partial}{\partial z} \left(\frac{\partial G}{\partial w^1}, \ldots, \frac{\partial G}{\partial w^n}\right) \Bigg|_{(0,0,0;\mu)} .$$

If only the bifurcation coefficients μ_2, τ_2 and β_2 are desired in a particular application of the Hopf Bifurcation Theorem, then one need compute nothing more, because the formula (5.9) for $c_1(0)$ involves just $g_{ij}(0)$ $(i + j = 2)$ and $g_{21}(0)$. If μ_4, τ_4 and β_4 are desired, then one must also compute the g_{ij} for $i + j = 3$, and g_{40}, g_{31}, g_{22}, g_{13}, and g_{32} . This calculation is sufficiently formidable that we again resorted to using the computer and a language for symbolic manipulation so as to decrease the chance for error in the results. The results are listed below. We note that the only unusual calculations involved are the calculations of w_{21} and w_{12} . In the formulae below expressions of the form $B^{INV}b$

represent the unique solution x of $Bx = b$ for which both

$$\langle \mathrm{Re}\, q^*, x\rangle = 0 \quad \text{and} \quad \langle \mathrm{Im}\, q^*, x\rangle = 0 ,$$

where it is known that $\langle \mathrm{Re}\, q^*, b\rangle = \langle \mathrm{Im}\, q^*, b\rangle = 0$ and that q^* and $\bar{q}^*$ are eigenvectors of B^H. The following formulae were first derived by Hassard and Wan [46]. In them G^2_{nm} is the Hessian matrix with elements

$$(G^2_{mn})_{k\ell} = \frac{\partial^2 G_{mn}}{\partial w^k \partial w^\ell} \qquad (k,\ell = 1,\ldots,n),$$

and the quantities H^1_{nm} and H^2_{00} are defined analogously to G^1_{nm} and G^2_{00} except that the components of the vector H^1_{nm} and the elements of the matrix H^2_{00} are understood to be n-dimensional column vectors.

$$w_{30} = (3\lambda - A)^{-1}[H_{30} + 3H^1_{10}w_{20} - 3G_{20}w_{20} - 3\bar{G}_{02}w_{11}] ,$$

$$w_{21} = (2\lambda + \bar{\lambda} - A)^{\mathrm{INV}}[H_{21} + 2H^1_{10}w_{11} + H^1_{01}w_{20} - (G_{20} + 2\bar{G}_{11})w_{11} - \bar{G}_{02}w_{02} - 2G_{11}w_{20}] ,$$

$$w_{12} = \bar{w}_{21} , \quad w_{03} = \bar{w}_{30} ;$$

$$g_{30} = G_{30} + 3G^1_{10}w_{20} ,$$

$$g_{12} = G_{12} + G^1_{10}w_{02} + 2G^1_{01}w_{11} ,$$

$$g_{03} = G_{03} + 3G^1_{01}w_{02} ,$$

$$g_{40} = G_{40} + 4G^1_{10}w_{30} + 6G^1_{20}w_{20} + 3w^T_{20}G^2_{00}w_{20} ,$$

$$g_{31} = G_{31} + 3G^1_{10}w_{21} + G^1_{01}w_{30} + 3G^1_{20}w_{11} + 3G^1_{11}w_{20} + 3w^T_{20}G^2_{00}w_{11} ,$$

$$\begin{aligned} g_{22} = {} & G_{22} + 2G^1_{10}w_{12} + 2G^1_{01}w_{21} + G^1_{20}w_{02} + 4G^1_{11}w_{11} + G^1_{02}w_{20} \\ & + w^T_{20}G^2_{00}w_{02} + 2w^T_{11}G^2_{00}w_{11} , \end{aligned}$$

$$g_{13} = G_{13} + G^1_{10}w_{03} + 3G^1_{01}w_{12} + 3G^1_{11}w_{02} + 3G^1_{02}w_{11} + 3w^T_{11}G^2_{00}w_{02} ,$$

$$g_{04} = G_{04} + 4G^1_{01}w_{03} + 6G^1_{02}w_{02} + 3w^T_{02}G^2_{00}w_{02} ,$$

$$\begin{aligned} g_{32} = {} & G_{32} + 3G^1_{10}w_{22} + 2G^1_{01}w_{31} + 3G^1_{20}w_{12} + 6G^1_{11}w_{21} + G^1_{02}w_{30} \\ & + G^1_{30}w_{02} + 6G^1_{21}w_{11} + 3G^1_{12}w_{20} + w^T_{30}G^2_{00}w_{02} + 6w^T_{21}G^2_{00}w_{11} \\ & + 3w^T_{12}G^2_{00}w_{20} + 3w^T_{20}G^2_{10}w_{02} + 6w^T_{11}G^2_{10}w_{11} + 6w^T_{20}G^2_{01}w_{11} ; \end{aligned}$$

$$\begin{aligned} w_{31} = {} & (3\lambda+\bar{\lambda}-A)^{-1}[H_{31}+3H^1_{20}w_{11} + 3H^1_{11}w_{20} + 3H^1_{10}w_{21} + H^1_{01}w_{30} \\ & + 3w^T_{20}H^2_{00}w_{11} - 3(2G^1_{10}w_{11} + G^1_{01}w_{20} + G_{21})w_{20} - (3G^1_{10}w_{20} \\ & + G_{30} + 3\bar{G}_{12} + 6\bar{G}^1_{01}w_{11} + 3\bar{G}^1_{10}w_{20})w_{11} - (3\bar{G}^1_{01}w_{20} + \bar{G}_{03})w_{02} \\ & - 3G_{11}w_{30} - 3(G_{20} + \bar{G}_{11})w_{21} - 3\bar{G}_{02}w_{12}] , \end{aligned}$$

$$\begin{aligned} w_{22} = {} & (4\mathrm{Re}\lambda-A)^{-1}[H_{22} + 2H^1_{10}w_{12}+2H^1_{01}w_{21}+w^T_{20}H^2_{00}w_{02} + 2w^T_{11}H^2_{00}w_{11} \\ & + H^1_{20}w_{02} + 4H^1_{11}w_{11} + H^1_{02}w_{20} - 2(G^1_{10}w_{02} + 2G^1_{01}w_{11} + G_{12})w_{20} \\ & -2(2G^1_{10}w_{11} + G^1_{01}w_{20} + G_{21})w_{11} - G_{02}w_{30} - 4G_{11}w_{21} - G_{20}w_{12} \\ & -2(2\bar{G}^1_{10}w_{11} + \bar{G}^1_{01}w_{02} + \bar{G}_{21})w_{11} - 2(\bar{G}^1_{10}w_{20} + 2\bar{G}^1_{01}w_{11} \\ & + \bar{G}_{12})w_{02} - \bar{G}_{20}w_{21} - 4\bar{G}_{11}w_{12} - \bar{G}_{02}w_{03}] , \end{aligned}$$

$$w_{40} = (4\lambda - A)^{-1}[H_{40} + 6H^1_{20}w_{20} + 4H^1_{10}w_{30} + 3w^T_{20}H^2_{00}w_{20}$$

$$- 4(3G^1_{10}w_{20} + G_{30})w_{20} - 6G_{20}w_{30} - 4(3\overline{G}^1_{01}w_{20} + \overline{G}_{03})w_{11}$$

$$- 6\overline{G}_{02}w_{21}] ,$$

$w_{13} = \overline{w}_{31}$, and $w_{04} = \overline{w}_{40}$.

Remark. In [46], Hassard and Wan performed a preliminary coordinate transformation to put the Jacobian matrix $A(\mu)$ into a real canonical form, as in Kelly's original proof of the Center Manifold Theorem. Indeed, this is the approach advocated in Chapter 2. The system corresponding to (6.6) in [46] has the form

$$\dot{z} = \lambda z + G(z,\overline{z},w;\mu)$$

$$\dot{w} = Dz + H(z,\overline{z},w;\mu) ,$$

where $w \in \mathbb{R}^{n-2}$ add the eigenvalues of D are $\lambda_3,\ldots,\lambda_n$. The operator corresponding to L in (6.15) is $\lambda z \frac{\partial}{\partial z} + \overline{\lambda}\overline{z}\frac{\partial}{\partial \overline{z}} - D$, so that all the matrices involved in expanding the center manifold are (classically) invertible. There is essentially no difference between the two approaches. The one we have taken in this section requires only q and q^* to produce the canonical form (6.6). The projection matrix P of [46, 57] may also be defined using only q and q^*: the first two columns of P are the real and imaginary parts of q and the remaining $n - 2$ columns of P may be formed from any set of $n - 2$ real, independent vectors v_j such that

$$\langle \mathrm{Re}\, q^*, v_j\rangle = \langle \mathrm{Im}\, q^*, v_j\rangle = 0 .$$

Remark. The formulae for g_{04} and w_{40} above are included for completeness: they are not actually used in computing $c_2(0)$ by means of (5.12) .

7. EXERCISES

Exercises 1 through 5 below illustrate Floquet theory, while Exercises 6 through 11 illustrate center manifold expansions. Exercise 11 uses the center manifold approach to develop a theory of stationary bifurcations.

1. Consider the 3-dimensional system

$$\dot{x}_1 = (\nu - 1)x_1 - x_2 + x_1 x_3 \equiv f^{(1)}(x;\nu)$$

$$\dot{x}_2 = x_1 + (\nu - 1)x_2 + x_2 x_3 \equiv f^{(2)}(x;\nu)$$

$$\dot{x}_3 = \nu x_3 - (x_1^2 + x_2^2 + x_3^2) \equiv f^{(3)}(x;\nu) .$$

Show that this system has a periodic solution for which x_3 is constant if $1/2 < \nu < 1$.

2. Let

$$p(t;\nu) = (R(\nu)\cos t \, , \, R(\nu) \sin t \, , \, 1 - \nu)$$

denote the periodic solution found in Exercise 1. Form the 2π-periodic matrix

$$A(t;\nu) = \left[\frac{\partial f^i}{\partial x_j} (p(t,\nu);\nu); \quad i,j = 1,2,3 \right] .$$

Find the general solution (with 3 arbitrary constants) of the variational system

$$\dot{y} = A(t;\nu)y .$$

(Directions: Rewrite the system first in terms of $z = y_1 + iy_2$. Then rewrite the system in terms of $\xi = \xi_1 + i\xi_2 \equiv e^{-it}z$. The system for ξ_1, ξ_2, y_3 will have constant coefficients.)

3. Write the solution $(y_1(t), y_2(t), y_3(t))^T$ from Exercise 2 above in the form

$$P(t;\nu)\begin{pmatrix}\xi_1\\ \xi_2\\ y_3\end{pmatrix},$$

where $P(t,\nu)$ is a 2π-periodic matrix. Hence represent the solution of the variational system in the form

$$y \equiv \begin{pmatrix}y_1\\ y_2\\ y_3\end{pmatrix} = P(t;\nu)e^{B(\nu)t}\begin{pmatrix}\xi_1(0)\\ \xi_2(0)\\ y_3(0)\end{pmatrix}.$$

4. Find the matrix C such that for any choice of initial conditions $y(0)$ for the system in Exercise 2,

$$y(2\pi) = C\,y(0).$$

Calculate the eigenvalues of C, i.e., the characteristic multipliers.
(Hints. Show that if v is an eigenvector for B corresponding to an eigenvalue λ, then v is an eigenvector of $e^{2\pi B}$ corresponding to the eigenvalue $e^{2\pi\lambda}$. Show that C is similar to $e^{2\pi B}$ and so has the same eigenvalues.) Describe all possible sets of characteristic exponents. (Note that the eigenvalues of B provide one set of characteristic exponents.)

5. For what values of ν $(1/2 < \nu < 1)$ is the periodic solution $p(t;\nu)$ obtained in Exercise 1 orbitally asymptotically stable with asymptotic phase? For what values of

ν $(1/2 < \nu < 1)$ is the periodic solution $p(t;\nu)$ unstable?

6. Consider the system

(a) $\dot{x} = xy$

$\dot{y} = -y + x^2$

in a neighborhood of the origin. Observe that the eigenvalues of the linear system

(b) $\dot{x} = 0$

$\dot{y} = -y$

corresponding to (a) at $(0,0)$ are 0 and -1. Draw the phase portrait of (b). Identify the stable manifold in this picture. Note that the center manifold for (b) consists of the x-axis, all of whose points are equilibria. By the Stable-Unstable-Center Manifold Theorem, the phase portrait of (a) is homeomorphically equivalent to that of (b) in a neighborhood of $(0,0)$. Attempt to sketch the phase portrait of (a). What is the stable manifold?
The center manifold for (a) has an expansion

(c) $y = ax + bx^2 + cx^2 + dx^4 + \dots$

near $(0,0)$. Differentiate (c) along trajectories of (a) and replace $\dot{x}$, $\dot{y}$, and y in the result by their series expansions in powers of x using (a) and (c). Equate the coefficients of corresponding powers of x on both sides of the resulting identity, and thus show that

$$a = 0 \ , \quad b = 1 \ , \quad c = 0 \ , \quad d = -2, \dots \ .$$

Using these coefficient values, substitute from (c) for y in the first of equations (a). The result is the projection of the system (a) onto its center manifold (which is unique).

Now complete your sketch of the phase portrait (a) near $(0,0)$ using the information you have computed. Note that the infinite series (c) diverges except at $x = 0$. However, near $x = 0$ partial sums represent the y-coordinate on the center manifold with an error that is of the order of magnitude of the first neglected term.

7. Consider the 3-dimensional system

$$\dot{x}_1 = \mu x_1 - x_2$$

$$\dot{x}_2 = x_1 + \mu x_2$$

$$\dot{x}_3 = -x_3 + x_1 x_2 ,$$

where μ is a real parameter. Let $z = x_1 + ix_2$, $\lambda = \mu + i$, and solve the system for arbitrary initial conditions $z(0), x_3(0)$. In each of the cases $\mu < 0$, $\mu = 0$, $\mu > 0$ describe the stable (resp. unstable) manifolds, i.e., the sets of initial conditions that produce solutions tending to the stationary solution in the limit $t \to +\infty$ (resp. $t \to -\infty$). For $\mu = 0$ describe the initial conditions producing periodic solutions.

8. Suppose that the system

$$\dot{x}_1 = \mu x_1 - x_2$$

$$(*) \qquad \dot{x}_2 = x_1 + \mu x_2$$

$$\dot{x}_3 = -x_3 + x_1 x_2$$

has solutions $(x_1, x_2, x_3)^T$ that satisfy

$$x_3 = w(z, \overline{z}; \mu) ,$$

where $z = x_1 + ix_2$ and w is smooth. Show that w satisfies the partial differential equation

$$w + \lambda z \frac{\partial w}{\partial z} + \bar{\lambda}\bar{z} \frac{\partial w}{\partial \bar{z}} = \frac{1}{4i}(z^2 - \bar{z}^2) ,$$

where $\lambda(\mu)$ is an eigenvalue of the system (*) linearized about the origin. Assume that the partial differential equation has solutions w of the form

$$w = \frac{w_{20}}{2} z^2 + w_{11} z\bar{z} + \frac{w_{02}}{2} \bar{z}^2 ,$$

and solve explicitly for the coefficients w_{20}, w_{11}, w_{02}. Obtain all solutions (z, x_3) of

$$\dot{z} = (\mu + i)z$$

$$\dot{x}_3 = -x_3 + \frac{z^2 - \bar{z}^2}{4i}$$

that satisfy

$$x_3(t) = w(z(t), \bar{z}(t); \mu) .$$

In each of the cases $\mu < 0$, $\mu = 0$, $\mu > 0$ describe the behaviour of these solutions.

<u>Remark</u>. For each fixed μ, the set

$$\{(x_1, x_2, x_3) \mid x_3 = w(z, \bar{z}; \mu) \text{ where } z = x_1 + i\, x_2\}$$

is a 2-dimensional manifold within R^3. These manifolds are slices μ = constant of a 3-dimensional manifold within R^4, the center manifold for the suspended system considered in the following exercise.

9. Consider the 4-dimensional system

$$\dot{y}_1 = y_4 y_1 - y_2$$

$$\dot{y}_2 = y_1 + y_4 y_2$$

$$\dot{y}_3 = -y_3 + y_1 y_2$$

$$\dot{y}_4 = 0 \ .$$

Let $z = y_1 + iy_2$, and suppose there is a smooth center manifold of the form

$$y_3 = W(z,\bar{z},y_4) = \sum_{n=2}^{3} \sum_{i+j+k=n} \frac{W_{ijk} z^i \bar{z}^j y_4^k}{i!j!k!} + O(4) \ ,$$

where $O(4)$ means terms of order 4 in $z,\bar{z}$ and y_4. First show that W satisfies the partial differential equation

$$W + iz \frac{\partial W}{\partial z} - i\bar{z} \frac{\partial W}{\partial \bar{z}} = \frac{1}{4i} (z^2 - \bar{z}^2) - y_3 (iz \frac{\partial W}{\partial z} - i\bar{z} \frac{\partial W}{\partial \bar{z}}) \ ,$$

then obtain explicit expressions for the coefficients W_{ijk} such that $i + j + k = 2,3$. Compare these coefficients W_{ijk} with the coefficients $w_{ij}(\mu)$ found in Exercise 8 above.

Remark. The system in this Exercise is called a suspended system, because it was obtained from the system in Exercise 8 by suspending (adding on, "below" the original system) the equation $\dot{y}_4 = 0$ so as to convert the parameter μ into a variable. The Center Manifold Theorem [43, 72] can be applied directly to the original system, but only in the case $\mu = 0$. When applied to the suspended system, the Center Manifold Theorem yields more. The Theorem captures within the manifold W all recurrent behaviour of the original system in a neighborhood of

$(x_1, x_2, x_3) = (0, 0, 0)$, not just for $\mu = 0$ but for all μ in some interval around 0 .

10. Suppose that the system (cf. Example 4, Chapter 2)

$$\dot{y}_1 = (2\nu - 1 + y_3)y_1 - y_2$$

$$\dot{y}_2 = y_1 + (2\nu - 1 + y_3)y_2$$

$$\dot{y}_3 = -\nu y_3 - (y_1^2 + y_2^2 + y_3^2)$$

has solutions y_1, y_2, y_3 that satisfy

$$y_3 = w(z,\bar{z};\nu) ,$$

where $z = y_1 + iy_2$ and w is smooth. Find the partial differential equation that w must satisfy (cf. Exercise 8), and solve explicitly for the coefficients w_{20}, w_{11}, w_{02} as in Exercise 8.

Remark. In the above exercises the Jacobian matrix at the stationary point is in real normal form. This form simplifies the computation but is not necessary. In general, the center manifold expansion will begin with quadratic terms (rather than linear terms) whenever the Jacobian matrix has the form

$$\begin{bmatrix} A_+ & 0 & 0 \\ 0 & A_0 & 0 \\ 0 & 0 & A_- \end{bmatrix},$$

where A_+ , A_0 , A_- are real square blocks whose spectra have the properties

$$\operatorname{Re}\sigma(A_+) > 0 , \quad \operatorname{Re}\sigma(A_0) = 0 , \quad \operatorname{Re}\sigma(A_-) < 0 .$$

In the case of Hopf bifurcation, this matrix is the Jacobian for the suspended system at $(x,\mu) = (0,0)$, A_+ is absent (if one is to have any hope of finding bifurcating periodic solutions that are stable), A_0 is 3×3 and A_- is $(n-2) \times (n-2)$. The additional dimension in A_0 arises from the equation $\dot{\mu} = 0$. In the case of Hopf bifurcation and when the slices μ = constant of the center manifold are expanded directly (cf. Exercises 8 and 10), the expansion will begin with quadratic terms whenever the Jacobian matrix of the original system at the stationary point is of the form

$$\begin{bmatrix} A & 0 \\ 0 & D \end{bmatrix},$$

where A is real and 2×2, $\sigma(A) = \{\alpha(\mu) \pm i\omega(\mu)\}$ $(\alpha(0) = 0,\ \alpha'(0) \neq 0,\ \omega(0) > 0)$, and $\operatorname{Re}\sigma(D) < 0$ for $|\mu|$ sufficiently small. When, further, A has the real canonical form

$$A = \begin{bmatrix} \alpha(\mu) & -\omega(\mu) \\ \omega(\mu) & \alpha(\mu) \end{bmatrix},$$

the complex substitution $z = y_1 + iy_2$ replaces the linear algebra associated with the matrix A by complex arithmetic involving $\lambda(\mu) = \alpha(\mu) + i\omega(\mu)$, which is a convenience.

11. Theory of stationary bifurcations.

Let $x_*(\nu)$ be a stationary point for the system

$$\dot{x} = f(x;\nu), \tag{1}$$

where $x \in \mathbb{R}^n$ $(n \geq 1)$, $\nu \in \mathbb{R}^1$, and $f \in C^5$ jointly in x,ν.

Suppose that $x_*(\nu)$ is C^5 for ν in some neighborhood of a critical value $\nu = \nu_c$ at which the eigenvalues $\lambda_1(\nu),\ldots,\lambda_n(\nu)$ of the Jacobian matrix $\partial f/\partial x(z_*(\nu);\nu)$ obey $\lambda_1(\nu_c) = 0$ and $\mathrm{Re}\,\lambda_j(\nu_c) < 0$ for the remaining indices j. In this situation, a stationary (static) bifurcation may occur at $\nu = \nu_c$ from x_* to another stationary point x^*. The center manifold derivation of the theory of stationary bifurcations is sketched in the following exercise.

(a) Show that there is a matrix $P(\nu)$ such that for all ν in some neighborhood of ν_c the system (1), under the change of variables

$$\begin{aligned} x &= x_*(\nu) + P(\nu)y \\ \nu &= \nu_c + \mu \end{aligned} \tag{2}$$

becomes

$$\dot{y} = F(y;\mu) , \tag{3}$$

where $F(y;\mu)$ is C^4 jointly in y,μ and where $\partial F/\partial y\,(0;\mu)$ has the form

$$\frac{\partial F}{\partial y}(0;\mu) = \begin{bmatrix} \alpha(\mu) & 0 \\ 0 & D(\mu) \end{bmatrix} , \tag{4}$$

in which $\alpha(\mu) = \lambda_1(\nu_c + \mu)$ while $D(\mu)$ is real and has s eigenvalues $\lambda_j(\nu_c + \mu)$, $(j = 2,\ldots,n)$.

(b) Let $u = y_1$, $w = (y_2,\ldots,y_n)^T$, and show that (3) may be written in the form

$$\dot{u} = \alpha(\mu)u + g(u,w;\mu)$$
$$\dot{w} = D(\mu)w + h(u,w;\mu) \;, \tag{5}$$

where $g, h = O(|u|^2 \;, |u||w|, |w|^2)$ uniformly for all sufficiently small μ, and the right-hand side of (5) is C^4 jointly in u, w, μ.

(c) Let $\tilde{u} = (u,\mu)^T$, and show that (5) may be written in the form

$$\dot{\tilde{u}} = \tilde{g}(\tilde{u},w)$$
$$\dot{w} = D(0)w + \tilde{h}(\tilde{u},w) \;, \tag{6}$$

where $\tilde{g} \;, \tilde{h} = O(|\tilde{u}|^2, |\tilde{u}||w|, |w|^2)$.

(d) Apply the Center Manifold Theorem [43, 72] to show that (6) has a center manifold

$$\mathcal{C} = \{(\tilde{u},w)^T \mid w = W(\tilde{u}), \; |\tilde{u}| < \delta\}$$

for some sufficiently small δ, where W is C^4 in $\tilde{u} = (u,\mu)^T$ and $W = O(|\tilde{u}|^2)$. Show further that $W(\tilde{u})$ satisfies the partial differential equation

$$\frac{\partial W}{\partial \tilde{u}} g(\tilde{u},W(\tilde{u})) = D(0)W(\tilde{u}) + \tilde{h}(\tilde{u},W(\tilde{u})) \tag{7}$$

(e) Let $W(u;\mu)$ represent a slice μ = constant of the center manifold $\mathcal{C}$. Show that (7) implies

$$W'(u;\mu)[\alpha(\mu)u + g(u,W(u;\mu);\mu)]$$
$$= D(\mu)W + h(u,W(u;\mu);\mu) \;, \tag{8}$$

where (') denotes differentiation with respect to u. Note that (8) arises directly from (5), under the formal assumption that (5) has solutions $(u(t), w(t))$, where

$w(t)$ is of the form $w(t) = W(u(t);\mu)$.

(f) Use the properties of $W(\tilde{u})$ from part (d) above to show that $W(u;\mu)$ and $W'(u;\mu)$ expand as

$$W(u;\mu) = \frac{1}{2} w_2 u^2 + \frac{1}{3!} w_3 u^3 + O(|u|^4) ,$$
$$W'(u;\mu) = w_2 u + \frac{1}{2} w_3 u^2 + O(|u|^3) . \tag{9}$$

(g) Substitute the expansions (9) into (8), and so obtain explicit algebraic formulae for the $(n - 1)$-dimensional vectors w_2 and w_3 in terms of the coefficients in the expansions

$$g(u,w;\mu) = \frac{1}{2} g_2 u^2 + u g_1^1 w + \frac{1}{3!} g_3 u^3 + O(|u|^4)$$
$$h(u,w;\mu) = \frac{1}{2} h_2 u^2 + u h_1^1 w + \frac{1}{3!} h_3 u^3 + O(|u|^4) , \tag{10}$$

where for simplicity w has already been taken to be $O(|u|^2)$. In (10), g_2 and g_3 are scalars, g_1^1 is an $(n-1)$-dimensional row vector, h_2 and h_3 are column vectors, and h_1^1 is a square matrix. These coefficients are partial derivatives at $(u,w) = (0,0)$ of the functions g and h defined in (b) above. Note that the linear algebraic systems for w_2 and w_3 have unique solutions whenever μ is sufficiently small.

(h) The function $G(u;\mu) \equiv g(u,W(u;\mu);\mu)$ is C^4 in u for fixed μ , and expands as

$$G(u;\mu) = \frac{1}{2} G_2(\mu)u^2 + \frac{1}{3!} G_3(\mu)u^3 + O(|u|^4) . \tag{11}$$

Use (9) and (10) to obtain explicit algebraic formulae for the coefficients G_2 and G_3 in (11) .

(i) Consider the first order differential equation

$$\dot{u} = \alpha(\mu)u + G(u;\mu) , \tag{12}$$

i.e., the restriction of the system (5) to the slice μ = constant of the center manifold $\mathcal{C}$. Define

$$\Phi(u;\mu) \equiv \begin{cases} \alpha(\mu) + G(u;\mu)/u & \text{if } u \neq 0 \\ \alpha(\mu) & \text{if } u = 0 , \end{cases} \tag{13}$$

and note that (12) may be written as

$$\dot{u} = u\Phi(u;\mu) .$$

Show that Φ is C^3 jointly in u, μ and expands as

$$\Phi(u,\mu) = \alpha(\mu) + \frac{1}{2} G_2(\mu)u + \frac{1}{3} G_3(\mu)u_2 + O(|u|^3) . \tag{15}$$

(j) Consider the algebraic equation

$$\Phi(u;\mu) = 0 . \tag{16}$$

Assume that

$$G_2(0) \neq 0 . \tag{17}$$

Apply the Implicit Function Theorem [14, 43] to show that, for all μ in some neighborhood of 0, equation (16) has a unique solution $u = u^*(\mu)$ with the property $u^*(0) = 0$. Show that $u^*(\mu)$ is C^3 in μ, and obtain explicit algebraic formulae for the coefficients u_1 and u_2 of the expansion

$$u^*(\mu) = u_1\mu + u_2\mu^2 + O(|\mu|^3) . \tag{18}$$

Show that both $u^*(\mu)$ and $u_*(\mu) \equiv 0$ are stationary solutions of the differential equation (12), and are distinct provided

$$[\alpha'(0)]^2 + [\alpha''(0)]^2 \neq 0 \ . \tag{19}$$

(k) Back substitute to show that

$$x^*(\nu) \equiv x_*(\nu) + P(\nu) \begin{pmatrix} u^*(\nu - \nu_c) \\ \\ W(u^*(\nu - \nu_c); \nu - \nu_c) \end{pmatrix} \tag{20}$$

is a stationary solution of (1) for all ν sufficiently close to ν_c , and that $x^*(\nu)$ expands as

$$x^*(\nu) = x_*(\nu) + P(\nu) \begin{pmatrix} u_1\mu + u_2\mu^2 \\ \\ \frac{1}{2}\, u_1^2\mu^2 w_2(0) \end{pmatrix} + O(|\mu|^3) \ , \tag{21}$$

where $\mu \equiv \nu - \nu_c$. Show that the stationary solutions x^*, x_* are distinct under conditions (17), (19).

(l) Substitute $u = u^*(\mu) + v$ in (12) and write the differential equation for v :

$$\dot{v} = \beta(\mu)v + H(v;\mu) \ , \tag{22}$$

where $\beta(\mu)$ is C^3 in μ , $\beta(0) = 0$, H is C^3 jointly in v and μ , and $H = O(|v|^2)$. Obtain explicit algebraic formulae for the coefficients β_1 , β_2 in the expansion

$$\beta(\mu) = \beta_1\mu + \beta_2\mu^2 + O(|\mu|^3) \ . \tag{23}$$

Hence show that, under conditions (17) and (19),

$$\text{sgn}\ \beta(\mu) = -\text{sgn}\ \alpha(\mu) \tag{24}$$

for all sufficiently small μ . This is the "principle of exchange of stabilities."

(m) In (j) above, replace condition (17) with the condition

$$\alpha'(0) \neq 0 , \tag{25}$$

and apply the Implicit Function Theorem to solve (16) for $\mu = \mu^*(u)$ rather than $u = u^*(\mu)$. Obtain explicit algebraic formulae for the coefficients μ_1 and μ_2 of the expansion

$$\mu^*(u) = \mu_1 u + \mu_2 u^2 + O(|u|^3) . \tag{26}$$

Back substitute to show that

$$x^* \equiv x_*(\nu_c + \mu^*(u)) + P(\nu_c + \mu^*(u)) \begin{pmatrix} u \\ W(u;\mu^*(u)) \end{pmatrix} \tag{27}$$

is a stationary solution of (1) for all sufficiently small u . Analyze the stability of the stationary solution (27) as in (l) above.

(n) The number

$$G_2(0) = g_2(0) = \frac{\partial^2}{\partial u^2} g(u,0;0)\Big|_{u=0}$$

plays an especially important role in the theory of stationary bifurcations. Let $A = \partial f/\partial x(x_*(\nu_c);\nu_c)$, and let $q,\tilde{q}$ denote (respectively) real row and column vectors such that

$$Aq = 0,\ \tilde{q}A = 0,\ \tilde{q}q = 1 .$$

Under the assumption that q is normalized so as to coincide with the first column of $P(\nu_c)$, show that

$$g(u,0;0) = \tilde{q}f(x_*(\nu_c) + qu;\nu_c) ,$$

so $G_2(0)$ may be computed by means of the difference quotient approximation

$$G_2(0) = \tilde{q}[f(x_*(\nu_c) + hq;\nu_c) + f(x_*(\nu_c) - hq;\nu_c)]/h^2 + O(|h|^2) .$$

For a discussion of how small h should be taken, see Appendix D. In addition to bifurcation points, the reader interested in families of solutions of $f(x;\nu) = 0$ should also know about <u>limit points</u> and <u>continuation methods</u>; see, for example, [28].

Partial Answers

(g) $w_2 = (2\alpha I - D)^{-1} h_2$, and

$$w_3 = (3\alpha I - D)^{-1} [h_3 + 3(g_2 I + h_1^1)w_2] ,$$

where I is the $(n - 1)$-dimensional identity matrix.

(h) $G_2 = g_2, \quad G_3 = g_3 + 3g_1^1 w_2$.

(j) $u_1 = -2\alpha'(0)/G_2(0)$,

$$u_2 = -[\alpha''(0) + G_2^1(0)u_1 - \frac{1}{3} G_3(0)u_1^2]/G_2(0) .$$

(l) $\beta_1 = u_1 G_2(0)/2 = -\alpha'(0)$,

$$\beta_2 = \frac{1}{2} u_2 G_2(0) + u_1[\frac{1}{2} G_2^1(0) + \frac{1}{3} G_3(0)u_1]$$

$$= -\frac{1}{2} \alpha''(0) + \frac{1}{3} G_3(0)u_1^2 .$$

(m) $\mu_1 = -G_2(0)/2\alpha'(0)$,

$$\mu_2 = -[\alpha''(0)\mu_1^2 + G_2^1(0)\mu_1 + \frac{2}{3} G_3(0)]/2\alpha'(0) .$$

CHAPTER 2. APPLICATIONS: ORDINARY DIFFERENTIAL EQUATIONS (BY HAND)

1. INTRODUCTION

In this Chapter we present a variety of applications of Hopf's Bifurcation Theorem to autonomous ordinary differential systems. The examples all belong to the class of problems for which only a "reasonable" amount of effort is required to obtain closed form analytic expressions for μ_2, τ_2, β_2. Masochists and symbolic manipulators can find harder examples which can still be worked out by the techniques used here — we have treated some of these in Chapter 3.

The examples we have chosen to present in this Chapter are: the mass-spring-belt system described in the Introduction to Chapter 1, van der Pol's equation, a model chemical reactor due to Lefever and Prigogine (the Brusselator), and a system with some fluid-dynamic ancestry due to W. F. Langford.

A Recipe-Summary

Evaluation of μ_2, τ_2, β_2 for systems of ordinary differential equations

1. Select the bifurcation parameter ν. (There may be much freedom in making this choice. One should hope to be guided by the real-world meaning of the system.) Let

$$\dot{x} = f(x;\nu) \qquad (x \in \mathbb{R}^n)$$

denote the system to be studied.

2. Locate $x_*(\nu)$, the stationary point of interest, calculate the eigenvalues of the Jacobian matrix

$$A(\nu) = \left\{ \frac{\partial f^i}{\partial x_j}(x_*(\nu);\nu) \qquad (i,j = 1,\ldots,n) \right\},$$

and order them according to

$$\operatorname{Re} \lambda_1 \geq \operatorname{Re} \lambda_2 \geq \ldots \geq \operatorname{Re} \lambda_n .$$

3. Find a value ν_c such that $\operatorname{Re} \lambda_1(\nu_c) = 0$. If
(a) λ_1 and λ_2 are a conjugate pair $(\lambda_1(\nu) = \bar{\lambda}_2(\nu))$ for ν in an open interval including ν_c,
(b) $\operatorname{Re} \lambda_1'(\nu_c) \neq 0$,
(c) $\operatorname{Im} \lambda_1(\nu_c) \neq 0$, and
(d) $\operatorname{Re} \lambda_j(\nu_c) < 0 \quad (j = 3,\ldots,n)$,
then a Hopf bifurcation occurs.

4. If $A(\nu_c)$ is in the form

$$\begin{bmatrix} 0 & -\omega_0 & \mathbf{0} \\ \omega_0 & 0 & \\ \mathbf{0} & & D \end{bmatrix}$$

where $\omega_0 = \operatorname{Im} \lambda_1(\nu_c) > 0$, let $P = I$, the identity matrix; and go to step 5. Otherwise form a matrix P as follows. Let $P = (\operatorname{Re} v_1, -\operatorname{Im} v_1, r_3,\ldots,r_n)$, where v_1 is the eigenvector of $A(\nu_c)$ corresponding to $\lambda_1(\nu_c) = i\omega_0$ and where $r_3,\ldots,r_n$ are any set of real n-vectors which span the union of the (generalized) eigenspaces for $\lambda_3,\ldots,\lambda_n$ at $\nu = \nu_c$. Normalize v_1 so that its first nonvanishing component is 1.

5. Perform the change of variables

$$x = x_*(\nu_c) + Py ,$$

and let $\dot{y} = F(y)$ denote the system for y. The Jacobian matrix $\partial F^i/\partial y_j(0)$ $(i,j = 1,\ldots,n)$ will have the real canonical form

$$\frac{\partial F}{\partial y}(0) = \begin{bmatrix} 0 & -\omega_0 & 0 \\ \omega_0 & 0 & \\ 0 & & D \end{bmatrix} .$$

6. Calculate the following quantities, all to be evaluated at $\nu = \nu_c$, $y = 0$.

$$g_{11} = \frac{1}{4}\left[\frac{\partial^2 F^1}{\partial y_1^2} + \frac{\partial^2 F^1}{\partial y_2^2} + i\left(\frac{\partial^2 F^2}{\partial y_1^2} + \frac{\partial^2 F^2}{\partial y_2^2}\right)\right]$$

$$g_{02} = \frac{1}{4}\left[\frac{\partial^2 F^1}{\partial y_1^2} - \frac{\partial^2 F^1}{\partial y_2^2} - 2\frac{\partial^2 F^2}{\partial y_1 \partial y_2}\right.$$

$$\left. + i\left(\frac{\partial^2 F^2}{\partial y_1^2} - \frac{\partial^2 F^2}{\partial y_2^2} + 2\frac{\partial^2 F^1}{\partial y_1 \partial y_2}\right)\right]$$

$$g_{20} = \frac{1}{4}\left[\frac{\partial^2 F^1}{\partial y_1^2} - \frac{\partial^2 F^1}{\partial y_2^2} + 2\,\frac{\partial^2 F^2}{\partial y_1 \partial y_2}\right.$$

$$\left. + i\left(\frac{\partial^2 F^2}{\partial y_1^2} - \frac{\partial^2 F^2}{\partial y_2^2} - 2\,\frac{\partial^2 F^1}{\partial y_1 \partial y_2}\right)\right]$$

$$G_{21} = \frac{1}{8}\left[\frac{\partial^3 F^1}{\partial y_1^3} + \frac{\partial^3 F^1}{\partial y_1 \partial y_2^2} + \frac{\partial^3 F^2}{\partial y_1^2 \partial y_2} + \frac{\partial^3 F^2}{\partial y_2^3}\right.$$

$$\left. + i\left(\frac{\partial^3 F^2}{\partial y_1^3} + \frac{\partial^3 F^2}{\partial y_1 \partial y_2^2} - \frac{\partial^3 F^1}{\partial y_1^2 \partial y_2} - \frac{\partial^3 F^1}{\partial y_2^3}\right)\right] .$$

7. If $n = 2$, let $g_{21} = G_{21}$; and go to step 8. If $n > 2$, calculate the following.

Let

$$h_{11}^{k-2} = \frac{1}{4}\left[\frac{\partial^2 F^k}{\partial y_1^2} + \frac{\partial^2 F^k}{\partial y_2^2}\right] \quad (k = 3,\ldots,n),$$

$$h_{20}^{k-2} = \frac{1}{4}\left[\frac{\partial^2 F^k}{\partial y_1^2} - \frac{\partial^2 F^k}{\partial y_2^2} - 2i\,\frac{\partial^2 F^k}{\partial y_1 \partial y_2}\right] \quad (k = 3,\ldots,n) .$$

Solve the linear systems

$$Dw_{11} = -h_{11}, \quad (D - 2i\omega_0 I)w_{20} = -h_{20}$$

for the $n - 2$ dimensional vectors w_{11}, w_{20}. The

matrix D is from step 5.

$$\text{Let}\quad G^{k-2}_{110} = \frac{1}{2}\left[\frac{\partial^2 F^1}{\partial y_1 \partial y_k} + \frac{\partial^2 F^2}{\partial y_2 \partial y_k} + i\left(\frac{\partial^2 F^2}{\partial y_1 \partial y_k} - \frac{\partial^2 F^1}{\partial y_2 \partial y_k}\right)\right]$$

$$G^{k-2}_{101} = \frac{1}{2}\left[\frac{\partial^2 F^1}{\partial y_1 \partial y_k} - \frac{\partial^2 F^2}{\partial y_2 \partial y_k} + i\left(\frac{\partial^2 F^1}{\partial y_2 \partial y_k} + \frac{\partial^2 F^2}{\partial y_1 \partial y_k}\right)\right],$$

and let

$$g_{21} = G_{21} + \sum_{k=1}^{n-2} (2\, G^k_{110} w^k_{11} + G^k_{101} w^k_{20}) \; .$$

8. Let

$$c_1(0) = \frac{i}{2\omega_0}\left[g_{20}g_{11} - 2|g_{11}|^2 - \tfrac{1}{3}|g_{02}|^2\right] + \frac{g_{21}}{2} \; ; \text{ then}$$

$$\mu_2 = -\text{Re}\; c_1(0)/\alpha'(0) \; ,$$

$$\tau_2 = -(\text{Im}\; c_1(0) + \mu_2\omega'(0))/\omega_0 \; , \text{ and}$$

$$\beta_2 = 2\; \text{Re}\; c_1(0) \; ,$$

where $\alpha'(0) = \text{Re}\;\lambda_1'(\nu_c)$, $\omega'(0) = \text{Im}\;\lambda_1'(\nu_c)$.

9. The period and characteristic exponent are:

$$T = \frac{2\pi}{\omega_0}(1 + \tau_2\varepsilon^2 + O(\varepsilon^4))$$

$$\beta = \beta_2\varepsilon^2 + O(\varepsilon^4) \; , \text{ where } \; \varepsilon^2 = \frac{\nu - \nu_c}{\mu_2} + O(\nu - \nu_c)^2 \; ,$$

(provided $\mu_2 \neq 0$) ; and the periodic solutions themselves are (except for an arbitrary phase angle)

$$x = x_*(\nu) + Py ,$$

where

$$y_1 = \text{Re } z , \quad y_2 = \text{Im } z ,$$

$$(y_3,\ldots,y_n)^T = w_{11}|z|^2 + \text{Re}(w_{20}z^2) + O(|z|^3)$$

and

$$z = \varepsilon e^{2\pi it/T} + \frac{i\varepsilon^2}{6\omega_0} [g_{02}e^{-4\pi it/T}$$

$$- 3g_{20}e^{4\pi it/T} + 6g_{11}] + O(\varepsilon^3) .$$

10. Have a friend check the calculations independently, or use numerical techniques (Chapter 3) to verify the results.

Remark 1. Exercise 8 at the end of this Chapter will guide the interested (or dubious) reader through the derivation of this Recipe-Summary from the results of Chapter 1.

Remark 2. If one or more of μ_2, τ_2, and β_2 is 0, then one may be interested in calculating μ_4, τ_4, and β_4. However, the hand calculation of μ_4, τ_4 and β_4 tends to be a tedious procedure, which we do not in general recommend. Steps 4 through 8 must be performed for arbitrary ν near ν_c, not just at $\nu = \nu_c$, so that $c_1(\mu)$ and hence $c_1'(0)$ may be obtained. Also, $c_2(0)$ must be found and this calculation involves 5th order partial derivatives of F at $\nu = \nu_c$, $y = 0$. Only for certain simple systems (such as the van der Pol example below) should hand calculation of μ_4, τ_4 and β_4 be attempted.

Remark 3. The procedure described above is an effective technique for analyzing the Hopf bifurcation when the system of ordinary differential equations is simple enough to treat by hand. The question, what is the 'best' technique, has no single answer because different classes of problems can have different properties which bestow selective advantage upon one technique or another. For example, if the linearized system simplifies greatly under Laplace transformation, then direct application of harmonic balancing [3] has advantages. As another example, note that although we advocate preliminary coordinate transformation in the present Chapter, in subsequent Chapters we generally avoid such transformations.

In support of the use of bifurcation formulae as opposed to direct application of a general theory, we note that the use of formulae takes advantage of the simplification performed in derivation of the formulae, which in many problems helps considerably. Also, the various general theories available (Center manifold-Poincaré Normal Form, Lyapunov-Schmidt, Lyapunov Functions-Poincaré Normal Form, Integral Averaging, Harmonic Balancing, Describing Functions, etc.) all necessarily produce exactly the same formulae, once the formulae have been appropriately compared. The use of bifurcation formulae may thus be considered as application of one's favorite theory, in simplified form.

In support of our particular recipe-summary for hand computations, we note that if the preliminary coordinate transformation is not employed, the analytic effort involved in the change of coordinates may not be eliminated but rather postponed to a later stage in the computation where it may further complicate matters. Similarly, our complex arithmetic may be replaced with real arithmetic, but with complex arithmetic the task is generally simpler.

<u>Example 1</u>. The mass-spring belt system.

The system

$$\frac{dx_1}{dt} = x_2$$

$$\frac{dx_2}{dt} = \frac{1}{m}(\mathfrak{F}(\nu - x_2) - kx_1 - cx_2)$$

was partially analyzed in the Introduction to Chapter 1. At the critical value ν_c, where $\mathfrak{F}'(\nu_c) = -c$, the eigenvalues of the system linearized about $x_*(\nu_c) = (\mathfrak{F}(\nu_c)/k, 0)$ are

$$\lambda_1 = \bar{\lambda}_2 = i\omega_0 ,$$

where

$$\omega_0 = \sqrt{k/m} ;$$

and the eigenvector corresponding to λ_1 is $v_1 = (1, i\omega_0)^T$. The matrix P is therefore

$$P = \begin{pmatrix} 1 & 0 \\ 0 & -\omega_0 \end{pmatrix}$$

and the variables y_1 and y_2 are given by

$$x_1 = \mathfrak{F}(\nu_c)/k + y_1 , \quad x_2 = -\omega_0 y_2 .$$

In terms of the y variables, the system becomes

$$\dot{y}_1 = -\omega_0 y_2$$

$$\dot{y}_2 = \omega_0 y_1 - \frac{c}{m} y_2 + \frac{1}{m\omega_0}[\mathfrak{F}(\nu_c) - \mathfrak{F}(\nu_c + \omega_0 y_2)]$$

or

$$\dot{y} = \mathfrak{F}(y;\nu_c) .$$

The only nonvanishing second and third order derivatives of F are $\frac{\partial^2 F^2}{\partial y_2^2} = -\frac{\omega_0}{m}\mathfrak{F}''(\nu_c)$ and $\frac{\partial^3 F^2}{\partial y_2^3} = -\frac{\omega_0^2}{m}\mathfrak{F}'''(\nu_c)$. Referring to the summary of the evaluation of μ_2, τ_2, β_2 , we compute

$$g_{11} = -g_{02} = -g_{20} = \frac{i}{4}\frac{\partial^2 F^2}{\partial y_2^2} = -\frac{i\omega_0}{4m}\mathfrak{F}''(\nu_c) ,$$

$$G_{21} = \frac{1}{8}\frac{\partial^3 F^2}{\partial y_2^3} = -\frac{\omega_0^2}{8m}\mathfrak{F}'''(\nu_c) , \text{ and}$$

$$c_1(0) = -\frac{2i}{3\omega_0}|g_{11}|^2 + \frac{1}{2}G_{21} = -\frac{i\omega_0}{24m^2}(\mathfrak{F}''(\nu_c))^2 - \frac{\omega_0^2}{16m}\mathfrak{F}'''(\nu_c) .$$

Thus

$$\mu_2 = -\text{Re } c_1(0)/\alpha'(0) = -\omega_0^2\mathfrak{F}'''(\nu_c)/(8\mathfrak{F}''(\nu_c)) > 0 ,$$

$$\beta_2 = 2\text{Re } c_1(0) = -\omega_0^2\mathfrak{F}'''(\nu_c)/(8m) > 0 , \text{ and}$$

$$\tau_2 = -(\text{Im } c_1(0) + \mu_2\omega'(0))/\omega_0 = (\mathfrak{F}''(\nu_c))^2/(24m^2) > 0 .$$

Since $\mu_2 > 0$, the periodic solutions exist for $\nu > \nu_c$ and since $\beta_2 > 0$, they are unstable. An approximation to the solutions is the expression

$$x_*(\nu_c) + \left[\frac{\nu - \nu_c}{\mu_2}\right]^{1/2} \text{Re } (e^{2\pi it/T}v_1) + O(\nu - \nu_c) ,$$

where

$$T = \frac{2\pi}{\omega_0}\left[1 + \tau_2\left[\frac{\nu - \nu_c}{\mu_2}\right] + O((\nu - \nu_c)^2)\right] ,$$

or

$$x(t) = \begin{bmatrix} \mathfrak{F}(\nu_c)/k \\ \\ 0 \end{bmatrix} + \left[\frac{8m\mathfrak{F}''(\nu_c)(\nu - \nu_c)}{k(-\mathfrak{F}'''(\nu_c))}\right]^{1/2} \begin{bmatrix} \cos(2\pi t/T) \\ \\ -\omega_0 \sin(2\pi t/T) \end{bmatrix}$$

$$+ O(\nu - \nu_c) ,$$

as given in Chapter 1. The interested reader may follow step 9 of the Recipe-Summary to evaluate the $O(\nu - \nu_c)$-term in the approximation for $x(t)$.

Example 2. van der Pol's equation.

The equations for the RLC electric circuit illustrated below

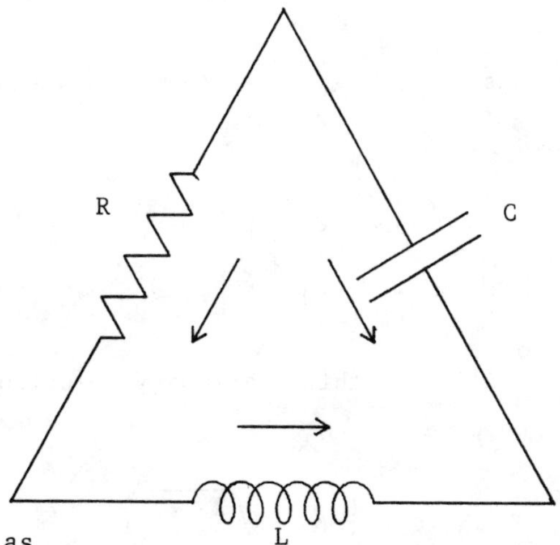

can be written as

$$i_C = C\,\frac{dv_C}{dt} , \quad V_L = L\,\frac{di_L}{dt} , \quad v_R = \phi(i_R) ,$$

$$i_R = i_L = -i_C , \quad v_R + v_L = v_C ,$$

where the i's are the currents in the branches indicated by the subscripts and where $v_R = \phi(i_R)$ is a generalized Ohm's law, characteristic of the "resistor" R, which is actually an active device. If we set $i_L = x$, $v_C = -(L/C)^{1/2} y$ and $t = (LC)^{1/2}\tau$, then the equations take the form

$$\dot{x} = -y - f(x)$$

$$\dot{y} = x$$

where $f(x) = (L/C)^{1/2} \phi(x)$ and '.' denotes differentiation with respect to the scaled time variable τ; for a full derivation and discussion of this system see Hirsch-Smale [51, Ch. 10]. Further, if the resistance is described by the function

$$f(x) = -\mu x + x^3 ,$$

then the system is a form of van der Pol's equation. The parameter μ controls the amount of "negative resistance" or "gain" of the device R.

For all values of μ, $(x,y) = (0,0)$ is a stationary point. Now

$$\begin{pmatrix} \dot{x} \\ \dot{y} \end{pmatrix} = \begin{pmatrix} +\mu & -1 \\ 1 & 0 \end{pmatrix} \begin{pmatrix} x \\ y \end{pmatrix} - \begin{pmatrix} x^3 \\ 0 \end{pmatrix} ,$$

so the linear stability of this stationary solution is determined by the eigenvalues

$$\lambda_{1,2} = \frac{1}{2} (\mu \pm \sqrt{\mu^2 - 4}) .$$

For $\mu \leq -2$ the eigenvalues are real and negative; for $-2 < \mu < 0$ they form a complex conjugate pair with negative real part; for $0 < \mu < 2$ they form a complex conjugate pair with positive real part; and for $\mu \geq 2$ they are real and positive.

Thus for fixed $\mu < 0$, Lyapunov's theorem [93] shows that $(0,0)$ is an asymptotically stable stationary point. (In fact, for $-1 < \mu < 0$, all solutions tend asymptotically to the steady state $(0,0)$ as $t \to \infty$, and $(0,0)$ is globally asymptotically stable; see Hirsch-Smale [51, Ch. 10].)

As μ is increased past $\mu = 0$, the stationary solution loses stability due to the complex conjugate pair of eigenvalues

$$\lambda_{1,2} = \alpha(\mu) \pm i\omega(\mu) ,$$

where $\alpha(\mu) = \mu/2$ and $\omega(\mu) = \sqrt{1 - \alpha^2}$. Moreover, $\alpha'(0) = 1/2 > 0$, so Hopf's theorem applies: there exists a family of periodic solutions bifurcating from $(0,0)$.

At $\mu = 0$, the Jacobian matrix of the system for (x,y) is already in real canonical form, and we may compute $c_1(0)$ (hence μ_2, τ_2, β_2) directly from the functions

$$F^1(x,y;0) = -y - x^3 , \quad F^2(x,y;0) = 0 .$$

The formulae give

$$g_{11} = g_{02} = g_{20} = 0 , \quad g_{21} = -3/4 , \quad c_1(0) = -3/8 ,$$

$$\mu_2 = 3/4 , \quad \tau_2 = 0 , \quad \text{and} \quad \beta_2 = -3/4 .$$

Since $\mu_2 > 0$, the periodic solutions exist for $\mu > 0$ and are stable. This information is consistent with the global analysis of Hirsch-Smale [51, Ch. 10] which shows that for each $0 < \mu \leq 1$ the van der Pol system has a unique periodic solution which is globally asymptotically orbitally stable as $t \to \infty$. Also since $\mu_2 \neq 0$, Theorem III implies that the period $T(\mu)$ and characteristic multiplier $\beta(\mu)$ are analytic in μ.

Since $\tau_2 = 0$ and since the higher order computation is not too difficult, we shall compute μ_4, τ_4 and β_4 as well.

To do this we need $c_1(\mu)$ and $c_2(0)$.

It is desirable to perform a coordinate transformation $P(\mu)$ such that for each μ near $\mu = 0$, the Jacobian matrix of the transformed system is in real canonical form.

The eigenvector of the matrix

$$\begin{bmatrix} \mu & -1 \\ 1 & 0 \end{bmatrix}$$

corresponding to $\lambda_1 = \alpha + i\omega$ is

$$v_1 = \begin{bmatrix} 1 \\ \alpha - i\omega \end{bmatrix},$$

so that

$$P = (\mathrm{Re}\ v_1, -\mathrm{Im}\ v_1) = \begin{bmatrix} 1 & 0 \\ \alpha & \omega \end{bmatrix}.$$

When we set

$$\begin{bmatrix} x \\ y \end{bmatrix} = P \begin{bmatrix} y_1 \\ y_2 \end{bmatrix},$$

the system for y_1, y_2 is

$$\dot{y}_1 = \alpha y_1 - \omega y_2 - y_1^3 = F^1(y_1, y_2; \mu)$$

$$\dot{y}_2 = \omega y_1 + \alpha y_2 + \frac{\alpha}{\omega} y_1^3 = F^2(y_1, y_2; \mu).$$

The Jacobian matrix for this system at $(y_1, y_2) = (0,0)$ is in real canonical form for any $\mu \varepsilon (-1,1)$ and, in particular, for μ near 0.

Let

$$z = y_1 + iy_2 .$$

Then the system for z is

$$\dot{z} = \lambda z + (\frac{i\alpha}{\omega} - 1)(\frac{z + \bar{z}}{2})^3$$

$$= \lambda z + \gamma(z^3 + 3z^2\bar{z} + 3z\bar{z}^2 + \bar{z}^3) ,$$

where

$$\gamma = \frac{1}{8}(\frac{i\alpha}{\omega} - 1) , \quad \lambda = \lambda_1 = \alpha(\mu) + i\omega(\mu) .$$

This system is

$$\dot{z} = \lambda z + g(z,\bar{z};\mu)$$

in the notation of Chapter 1. Hence

$$g_{ij} = 0 \quad \text{if} \quad i + j \neq 3$$

$$g_{30} = g_{21} = g_{12} = g_{03} = 6\gamma .$$

The formulae from Chapter 1 then imply:

$$c_1(\mu) = 3\gamma , \quad c_1(0) = -3/8 , \quad c_1'(0) = i/16 ,$$

$$\mu_2 = 3/4 , \quad \tau_2 = 0 , \quad \beta_2 = -3/4 ,$$

$$\chi_{ij} = 0 \quad \text{for} \quad i + j = 2 \quad \text{or} \quad 4 ,$$

$$\chi_{30} = 3i/8 , \quad \chi_{12} = -3i/8 , \quad \chi_{03} = -3i/16 ,$$

$$c_2(0) = -27i/256 ,$$

$$\mu_4 = 0\ , \quad \tau_4 = 9/256\ , \quad \beta_4 = 0\ .$$

The functions $T(\mu)$ and $\beta(\mu)$ therefore have the expansions

$$T(\mu) = 2\pi(1 + \frac{\mu^2}{16} + O(\mu^3))\ ,$$

$$\beta(\mu) = -\mu + O(\mu^3)\ ,$$

and the periodic solutions themselves are approximated by

$$p(t;\mu) = (y_1, \alpha y_1 + \omega y_2)^T\ ,$$

where

$$\alpha = \mu/2\ , \quad \omega = \sqrt{1 - \alpha^2}\ , \text{ and } \quad y_1 + iy_2 = z$$

with

$$z = \xi + \frac{i}{16}\xi^3 - \frac{3i}{16}\xi\bar{\xi}^2 - \frac{i}{32}\bar{\xi}^3 + O(\varepsilon^5)\ ,$$

$$\xi = \varepsilon e^{i2\pi t/T(\mu)}\ ,$$

and

$$\varepsilon^2 = \frac{4}{3}\mu + O(\mu^3)\ .$$

The following table compares these approximate results with the "exact" results obtained by a simple shooting scheme. Each shot (numerical integration) began on the positive x axis, and the initial position x_0 was varied until a trajectory was generated that returned to the initial position.

TABLE 2.1

μ	$T(\mu)$	$2\pi(1+\mu^2/16)$	$x(y=0)$	$(4\mu/3)^{1/2}$
.025	6.28343	6.28343	.18257	.18257
.05	6.28417	6.28417	.25817	.25820
.1	6.28711	6.28711	.36496	.36515
.2	6.29888	6.29889	.51537	.51640
.4	6.34574	6.34602	.72462	.73030

The column $x(y=0)$ contains the intercepts with the positive x-axis of the periodic solutions so computed.

The agreement between the numerical results and the approximate analytical results is of the sort expected; namely, it is better for smaller values of μ. The numerical results support the correctness of our calculated values for the bifurcation parameters.

Example 3. Bulk oscillations of the Brusselator.

The pair of nonlinear diffusion equations

$$\frac{\partial X}{\partial t} = D_1 \Delta X - (B + 1)X + X^2Y + A$$

$$\frac{\partial Y}{\partial t} = D_2 \Delta Y + BX - X^2Y$$

was posed by Lefever and Prigogine [75] in 1968 as a model system for an autocatalytic chemical reaction with diffusion. Here A and B are concentrations of certain "initial" substances and are assumed to be constant. The unknowns X and Y are concentrations of two intermediates, and D_1 and D_2 are their respective diffusion coefficients. The symbol Δ represents the Laplacian in the appropriate number of space variables, and the term X^2Y represents the autocatalytic step. We shall return to this system in Chapters 3 and 5. Here we shall assume, however, that X and Y are space independent; consequently, the reaction is governed by the ordinary differential system

$$\dot{X} = -(B + 1)X + X^2Y + A$$

$$\dot{Y} = BX - X^2Y .$$

The only stationary point of this system is

$$X = A \ , \quad Y = B/A \qquad (A,B > 0) \ ,$$

and the Jacobian matrix of this system at this equilibrium is

$$\begin{bmatrix} B - 1 & A^2 \\ -B & -A^2 \end{bmatrix} .$$

The eigenvalues λ obey the characteristic equation

$$\lambda^2 - (B - 1 - A^2)\lambda + A^2 = 0 .$$

Let

$$\alpha = \frac{1}{2} [B - (1 + A^2)] .$$

Then if $\alpha^2 < A^2$, the roots λ form a complex conjugate pair

$$\lambda_1 = \bar{\lambda}_2 = \alpha + i\omega ,$$

where

$$\omega = \sqrt{A^2 - \alpha^2} .$$

We now choose B as the bifurcation parameter, and we note that as B is increased past $B_0 = 1 + A^2$, the stationary point $(A,B/A)$ loses linear stability since the complex conjugate pair of eigenvalues then has positive real part.

The eigenvector v_1 corresponding to λ_1 is

$$v_1 = \begin{pmatrix} 1 \\ (\alpha + 1 - B + i\omega)/A^2 \end{pmatrix} .$$

We define

$$P = (\mathrm{Re}\ v_1, -\mathrm{Im}\ v_1) = \begin{pmatrix} 1 & 0 \\ \dfrac{\alpha + 1 - B}{A^2} & \dfrac{-\omega}{A^2} \end{pmatrix} .$$

and

$$\begin{pmatrix} y_1 \\ y_2 \end{pmatrix} = P^{-1} \begin{pmatrix} X \\ Y \end{pmatrix} - P^{-1} \begin{pmatrix} A \\ B/A \end{pmatrix} .$$

Then the system for y_1, y_2 is

$$\dot{y}_1 = \alpha y_1 - \omega y_2 + h(y_1, [(\alpha + 1 - B)y_1 - \omega y_2]/A^2)$$

$$\dot{y}_2 = \omega y_1 + \alpha y_2 + \gamma h(y_1, [(\alpha + 1 - B)y_1 - \omega y_2]/A^2) ,$$

where

$$h(x,y) = BA^{-1} x^2 + (x + 2A) xy$$

and

$$\gamma = \frac{\alpha + 1 - B + A^2}{\omega} .$$

The bifurcation formulae we have previously derived apply immediately to this system. Since we shall evaluate only μ_2, τ_2, and β_2, which are obtained from $c_1(0)$ alone, we may set

$$B = B_0 = 1 + A^2 ,$$

so that in the above

$$\alpha = 0 \quad \text{and} \quad \omega = \omega_0 = A .$$

Thus the system becomes

$$\dot{y}_1 = -Ay_2 + F^1(y_1,y_2;0)$$

$$\dot{y}_2 = Ay_1 + F^2(y_1,y_2;0) ,$$

where

$$F^1(y_1,y_2;0) = (\frac{1}{A} - A)y_1^2 - 2y_1y_2 - y_1^3 - \frac{y_1^2y_2}{A} ,$$

and

$$F^2(y_1,y_2;0) = 0 .$$

Substituting in our formulae, we obtain

$$g_{11} = \frac{1}{2}(\frac{1}{A} - A) , \quad g_{02} = g_{11} - i , \quad g_{20} = g_{11} + i ,$$

$$g_{21} = \frac{1}{4}[-3 + i/A] ,$$

and

$$c_1(0) = \frac{i}{2A}\left[g_{20}g_{11} - 2|g_{11}|^2 - \frac{1}{3}|g_{02}|^2\right] + \frac{g_{21}}{2}$$

$$= -\left\{\frac{1}{4A^2} + \frac{1}{8} + i\left[\frac{1}{6A}(\frac{1}{A} - A)^2 + \frac{1}{24A}\right]\right\}.$$

Now

$$\alpha'(0) = 1/2 \quad \text{and} \quad \omega'(0) = 0 \text{ , so that}$$

$$\mu_2 = -\text{Re } c_1(0)/\alpha'(0)$$

$$= \frac{1}{2A^2} + \frac{1}{4} > 0 \text{ ,}$$

and

$$\tau_2 = -[\text{Im } c_1(0) + \mu_2\omega'(0)]/\omega(0)$$

$$= \frac{1}{6A^2}\left[(\frac{1}{A} - A)^2 + \frac{1}{4}\right] > 0 \text{ .}$$

Thus the bifurcation is always supercritical. Since

$$\beta_2 = -2\alpha'(0)\mu_2 < 0 \text{ ,}$$

the periodic solutions that arise are stable. Also, for B close enough to B_0, the period grows with increasing B :

$$T = \frac{2\pi}{A}\left\{1 + \frac{1}{3}\left[(\frac{1}{A} - A)^2 + \frac{1}{4}\right]\left[1 + \frac{A^2}{2}\right]^{-1}(B - B_0) + O(B - B_0)^2\right\} \text{ ,}$$

and (by Theorem III) is analytic in $B - B_0$ (or B). The periodic solutions themselves are approximated by

$$\begin{pmatrix} X \\ Y \end{pmatrix} = \begin{pmatrix} A \\ B_0/A \end{pmatrix} + \begin{pmatrix} 1 & 0 \\ -1 & -1/A \end{pmatrix}\begin{pmatrix} y_1 \\ y_2 \end{pmatrix} \text{ ,}$$

where

$$y_1 + iy_2 = [2A^2(1 + A^2/2)^{-1}(B - B_0)]^{1/2} \exp[2\pi i(t + \phi)/T]$$

$$+ O(B - B_0)$$

and ϕ is an arbitrary phase angle.

(The Hopf algorithm employed by Auchmuty and Nicholis [7] does not provide fully simplified expressions for their quantities γ_2 and ω_2. Our expressions for μ_2 and τ_2 are in fully simplified form as they arise from the Bifurcation Formulae.)

The periodic solutions of the ordinary differential system in this example are also periodic solutions of the original partial differential system. They are called bulk oscillations, because they are space independent. If the Brusselator with diffusion is considered on an interval and no flux (Neumann) boundary conditions are imposed at the endpoints of the interval, then these bulk oscillations are also stable as solutions of the partial differential system. This stability will be shown in Chapter 5. It is an application of center manifold theory for flows.

<u>Example</u> 4. Langford's System.

W. F. Langford (private communication) recently introduced a 3rd order system of ordinary differential equations, rich in bifurcational behaviour, by truncating an infinite system of ordinary differential equations, originally suggested by E. Hopf [56] as a possible model for fluid dynamic turbulence. Langford's system is deceptively simple; it is

$$\dot{x}_1 = (\nu - 1)x_1 - x_2 + x_1x_3$$

$$\dot{x}_2 = x_1 + (\nu - 1)x_2 + x_2x_3$$

$$\dot{x}_3 = \nu x_3 - (x_1^2 + x_2^2 + x_3^2) .$$

The only stationary solutions are

$$x_*^0 = (0,0,0)^T \text{ and } x_*^1 = (0,0,\nu)^T .$$

Linearized about x_*^0 the system has coefficient matrix

$$\begin{pmatrix} \nu - 1 & -1 & 0 \\ 1 & \nu - 1 & 0 \\ 0 & 0 & \nu \end{pmatrix} ,$$

whose eigenvalues are

$$\lambda_1^0 = \nu , \quad \lambda_{2,3}^0 = \nu - 1 \pm i .$$

Thus x_*^0 is linearly stable for $\nu < 0$ and linearly unstable for $\nu > 0$. (For any fixed $\nu < 0$, Lyapunov's theorem [93] applies and shows that x_*^0 is asymptotically stable as a solution of the nonlinear system.)

Linearized about x_*^1 the system has coefficient matrix

$$\begin{pmatrix} 2\nu - 1 & -1 & 0 \\ 1 & 2\nu - 1 & 0 \\ 0 & 0 & -\nu \end{pmatrix} ,$$

whose eigenvalues are

$$\lambda_{1,2}^1 = 2\nu - 1 \pm i , \quad \lambda_3^1 = -\nu .$$

Thus x_*^1 is linearly stable for $0 < \nu < 1/2$, but x_*^1 is linearly unstable for $\nu < 0$ or $\nu > 1/2$.

As ν is increased past 0 the stability lost by x_*^0 is gained by x_*^1 in a stationary bifurcation due to a single real eigenvalue (the other two are complex with negative real parts).

As ν is increased past 1/2, however, the stationary point x_*^1 loses stability due to a pair of pure imaginary eigenvalues $\lambda_{1,2}^1(\frac{1}{2}) = \pm i$; and a Hopf bifurcation from stationary to periodic solutions then results.

Let

$$x = x_*^1 + y .$$

The nonlinear system becomes

$$\dot{y}_1 = (2\nu - 1)y_1 - y_2 + y_1y_3$$

$$\dot{y}_2 = y_1 + (2\nu - 1)y_2 + y_2y_3$$

$$\dot{y}_3 = - \nu y_3 - (y_1^2 + y_2^2 + y_3^2) .$$

The bifurcation formulae apply directly to this system, since its Jacobian matrix at $y = 0$ is in real canonical form. Applying the formulae, we compute

$$g_{11} = g_{02} = g_{20} = G_{21} = 0 ,$$

$$h_{11} = -1 , \quad h_{20} = 0 ,$$

$$w_{11} = -2 , \quad w_{20} = 0 ,$$

$$g_{110}^1 = 1 , \quad g_{101}^1 = 0 , \quad g_{21} = 4 ,$$

$$c_1(0) = -2 , \quad \mu_2 = 1 , \quad \tau_2 = 0 , \quad \text{and} \quad \beta_2 = -4 .$$

Thus the periodic solutions $p(t;\nu)$ exist for $\nu > 1/2$, are asymptotically orbitally stable, and are approximated by

$$p(t;\nu) = x_*^1(\tfrac{1}{2}) + (\nu - \tfrac{1}{2})^{\frac{1}{2}} \operatorname{Re}\left(e^{it}\begin{bmatrix}1\\-i\\0\end{bmatrix}\right) + O(\nu - \tfrac{1}{2}) ,$$

$$= \begin{bmatrix}0\\0\\\frac{1}{2}\end{bmatrix} + (\nu - \tfrac{1}{2})^{\frac{1}{2}} \begin{bmatrix}\cos t\\\sin t\\0\end{bmatrix} + O(\nu - \tfrac{1}{2}) .$$

In the case of this example we can check the results analytically. In cylindrical coordinates $x_1 = r\cos\theta$, $x_2 = r\sin\theta$, $x_3 = x_3$ the system becomes

$$\dot{r} = (\nu - 1)r + rx_3$$

$$\dot{\theta} = 1$$

$$\dot{x}_3 = \nu x_3 - (r^2 + x_3^2)$$

and by inspection, we find the (exact) solution

$$r = R(\nu)$$

$$\theta = t$$

$$x_3 = 1 - \nu$$

where

$$R(\nu) = [(1 - \nu)(2\nu - 1)]^{1/2} = (2 - 2\nu)^{1/2}(\nu - 1/2)^{1/2}$$

$$= (\nu - 1/2)^{1/2} + O[(\nu - 1/2)]^{3/2} .$$

In Cartesian coordinates, this solution is

$$p(t;\nu) = \begin{pmatrix} 0 \\ 0 \\ 1-\nu \end{pmatrix} + R(\nu) \begin{bmatrix} \cos t \\ \sin t \\ 0 \end{bmatrix},$$

and is the one whose approximation we derived above. Expanding $R(\nu)$ about $\nu = 1/2$, we see that $\mu_2 = 1$ is correct. The period T is a constant, namely, 2π; so $\tau_2 = 0$ is also correct.

The periodic solution corresponds to the stationary solution $r = R(\nu)$, $x_3 = 1 - \nu$ of the 2-dimensional autonomous system

$$\begin{aligned} \dot{r} &= (\nu - 1 + x_3)r \\ \dot{x}_3 &= \nu x_3 - (r^2 + x_3^2) . \end{aligned} \qquad (*)$$

The Jacobian of this system at $r = R$, $x_3 = 1 - \nu$ is

$$\begin{bmatrix} 0 & R(\nu) \\ -2R(\nu) & 3\nu - 2 \end{bmatrix}$$

with characteristic polynomial

$$\lambda^2 - (3\nu - 2)\lambda + 2R^2(\nu)$$

and eigenvalues

$$\lambda_{1,2} = \frac{1}{2}\{3\nu - 2 \pm [25\nu^2 - 36\nu + 12]^{1/2}\} .$$

These eigenvalues are also the nonvanishing characteristic (Floquet) exponents associated with $p(t;\nu)$. In particular, the Floquet exponent whose expansion is

$$\beta(\nu) = \beta_2(\nu - 1/2) + O(\nu - 1/2)^2$$

is exactly

$$\beta(\nu) = \lambda_1(\nu) ,$$

as given above. Expanding $\lambda_1(\nu)$ about $\nu = 1/2$, we confirm that our previous evaluation

$$\beta_2 = -4$$

is correct.

We leave it as an exercise for the reader to show that at $\nu = 1$ there is a subcritical bifurcation from the stationary point x_*^0 to the same periodic solution $p(t;\nu)$, except that for $\nu \approx 1$, $\nu < 1$, $p(t;\nu)$ is unstable by virtue of two positive characteristic exponents.

There is even more bifurcational behavior in Langford's system, which we shall now pursue. The eigenvalues $\lambda_{1,2}$ above may be rewritten as

$$\lambda_{1,2} = \frac{1}{2} \{(3\nu - 2) \pm 5[(\nu - \nu_1)(\nu - \nu_2)]^{1/2}\} ,$$

where

$$\nu_1 = \frac{18 - 2\sqrt{6}}{25} , \quad \nu_2 = \frac{18 + 2\sqrt{6}}{25} .$$

Since $1/2 < \nu_1 < 2/3 < \nu_2 < 1$, we see that the stationary point $r = R$, $x_3 = 1 - \nu$ (equivalently, the periodic solution $p(t;\nu)$) is stable for $1/2 < \nu < 2/3$, with the eigenvalues $\lambda_{1,2}$ being real and negative for $1/2 < \nu \leq \nu_1$ but forming a complex conjugate pair with negative real part for $\nu_1 < \nu < 2/3$. At $\nu = 2/3$ the stationary point loses linear stability. For $2/3 < \nu < \nu_2$, the eigenvalues form a complex conjugate pair with positive real part; and for $\nu_2 \leq \nu < 1$ the eigenvalues are both real and positive. Thus a Hopf bifurcation for the 2-dimensional system occurs at $\nu = 2/3$. The resulting family of periodic solutions in the (r,x_3) variables occurs "on top of" the basic periodic solution, and so represents a family of bifurcating tori in the original coordinate space. The general theory of bifurcation to tori is beyond the scope of these Notes [60, 74, 97]. However, because of special symmetries in Langford's system, Hopf Bifurcation Theory is adequate to describe this phenomenon in the present case.

For ν near $2/3$ the eigenvector v_1 corresponding to the eigenvalue λ_1 associated with (*) is

$$v_1 = \begin{pmatrix} 1 \\ \lambda_1/R(\nu) \end{pmatrix} = \begin{pmatrix} 1 \\ [\alpha(\nu) + i\omega(\nu)]/R(\nu) \end{pmatrix} ,$$

where

$$\alpha(\nu) = (3\nu - 2)/2 \quad \text{and} \quad \omega(\nu) = \frac{1}{2}[36\nu - 25\nu^2 - 12]^{1/2} .$$

(Note: These α and ω are not the α and ω of the previous bifurcation calculation.) We next define

$$P_1 = (\text{Re } v_1, -\text{Im } v_1) = \begin{pmatrix} 1 & 0 \\ \frac{\alpha}{R} & \frac{-\omega}{R} \end{pmatrix} ,$$

and we define

$$\begin{pmatrix} y_1 \\ y_2 \end{pmatrix} = P_1^{-1} \begin{pmatrix} r \\ x_3 \end{pmatrix} - P_1^{-1} \begin{pmatrix} R \\ 1 - \nu \end{pmatrix} .$$

Again, since we shall only compute μ_2, τ_2 and β_2, it is enough to set $\nu = 2/3$ throughout the calculation, in which case

$$\alpha = 0 \ , \quad R = 1/3 \ , \quad \omega_0 = \frac{1}{3}\sqrt{2} \ ,$$

$$y_1 = r - \frac{1}{3} \ , \quad y_2 = -\frac{1}{3\omega_0}(x_3 - \frac{1}{3}) \ .$$

In the new variables y_1, y_2 the system (*) becomes

$$(**) \qquad \begin{aligned} \dot{y}_1 &= -\omega_0 y_2 - 3\omega_0 y_1 y_2 \\ \dot{y}_2 &= \omega_0 y_1 + \frac{1}{3\omega_0} y_1^2 + 3\omega_0 y_2^2 \ , \end{aligned}$$

Applying our bifurcation formulae in the Recipe-Summary, we compute:

$$g_{11} = 3i\sqrt{2}/4 \ , \quad g_{02} = -g_{11} \ , \quad g_{20} = i\sqrt{2}/4 \ ,$$

$$g_{21} = 0 \ , \quad c_1(0) = -9i/2\sqrt{2} \ , \quad \alpha'(0) = 3/2 \ , \quad \omega'(0) = 1/\sqrt{2}$$

$$\mu_2 = 0 \ , \quad \tau_2 = 27/4 \ , \quad \beta_2 = 0 \ .$$

Unfortunately, the result $\mu_2 = 0$ does not indicate the direction of bifurcation. We do know, however, that the periods of the oscillations increase as their amplitudes grow. The first impulse when encountering a case $\mu_2 = 0$ is to compute μ_4. In this case the impulse should be resisted, for a moment's thought shows that the periodic solutions all exist for $\mu = 0$, i.e., $\nu = 2/3$, so that μ_4 must also vanish a priori. To see this use symmetry. For all sufficiently small ε, if we pose the initial conditions $y_1(0) = \varepsilon$, $y_2(0) = 0$, the solution of (**) must exist for at least $2\pi/\omega_0$ units of time and must cross the line $y_2 = 0$ for some time near π/ω_0. Because of the symmetry of the y_1, y_2 phase plane, the trajectory backwards in time from the same initial conditions is the reflection in the line $y_2 = 0$ of the forwards trajectory, and the trajectories meet at $y_2 = 0$, $y_1 = -\varepsilon + O(\varepsilon^2)$. Thus there is a family of periodic solutions at $\nu = 2/3$, which is necessarily the family predicted by Theorem I.

This family of periodic solutions is described by

$$\begin{pmatrix} y_1(t;\varepsilon) \\ \\ y_2(t;\varepsilon) \end{pmatrix} = \varepsilon \begin{pmatrix} \cos(2\pi t/T_\varepsilon) \\ \\ -\sqrt{2}\sin(2\pi t/T_\varepsilon) \end{pmatrix} + O(\varepsilon^2) ,$$

where the period is

$$T_\varepsilon = \frac{2\pi}{\omega_0}(1 + \frac{27}{4}\varepsilon^2 + O(\varepsilon^4)) ,$$

and both Floquet exponents vanish.

In cylindrical coordinates the family of bifurcating tori, belonging to the system (*), is described by

$$r = \frac{1}{3} + \varepsilon \cos\left[\left(\frac{2\pi t}{T_\varepsilon}\right) + \phi_1\right] + O(\varepsilon^2)$$

$$x_3 = \frac{1}{3} + 2\varepsilon \sin\left[\left(\frac{2\pi t}{T_\varepsilon}\right) + \phi_1\right] + O(\varepsilon^2)$$

$$\theta = t + \phi_2 ,$$

where ϕ_1 and ϕ_2 are arbitrary phase angles. When Fourier analyzed, the family of tori display two discrete basic frequencies:

$$\omega_\varepsilon = 1 \text{ and } \omega_\varepsilon = \omega_0(1 - \frac{27}{4}\varepsilon^2 + O(\varepsilon^4)) .$$

Two members of this family of bifurcating tori, obtained by direct numerical integration, are shown in Figure 2.1 below.

Before leaving this example, we emphasize that the general

problem of bifurcation to tori is much more difficult than the example just studied might suggest. Only for certain systems possessing a high degree of symmetry will the technique of "Hopf upon Hopf" used above be applicable, in which case the resulting tori are necessarily quasiperiodic.

TWO MEMBERS OF FAMILY OF BIFURCATING
QUASIPERIODIC TORI

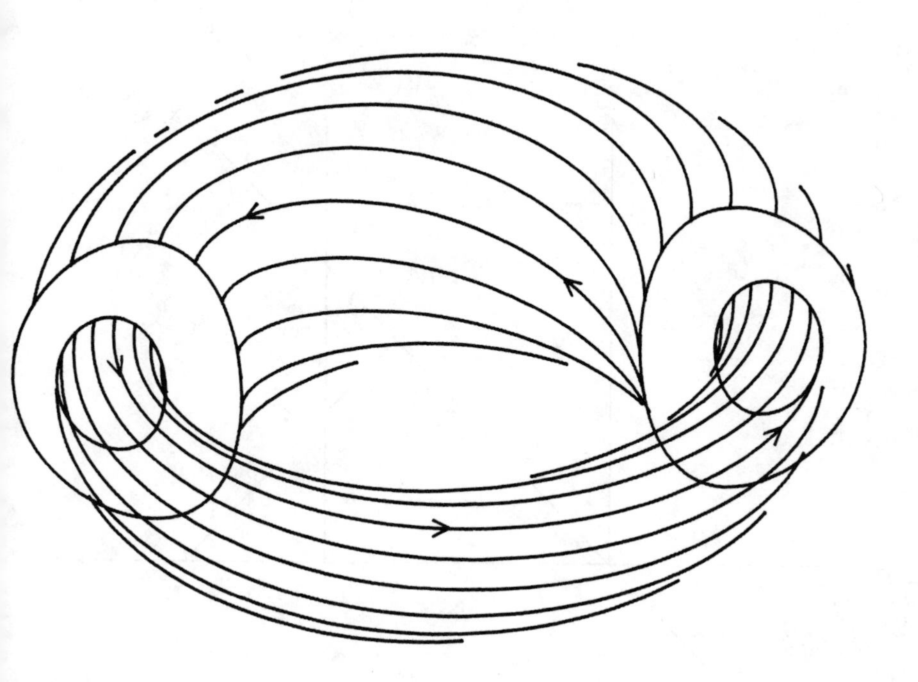

System: $\dot{x}_1 = (\nu - 1 + x_3)x_1 - x_2$

$\dot{x}_2 = x_1 + (\nu - 1 + x_3)x_2$

$\dot{x}_2 = \nu x_3 - (x_1^2 + x_2^2 + x_3^2)$;

the family occurs for $\nu \equiv \nu_c \equiv 2/3$.

$(\mu_2 = 0!)$

<u>Figure 2.1</u>

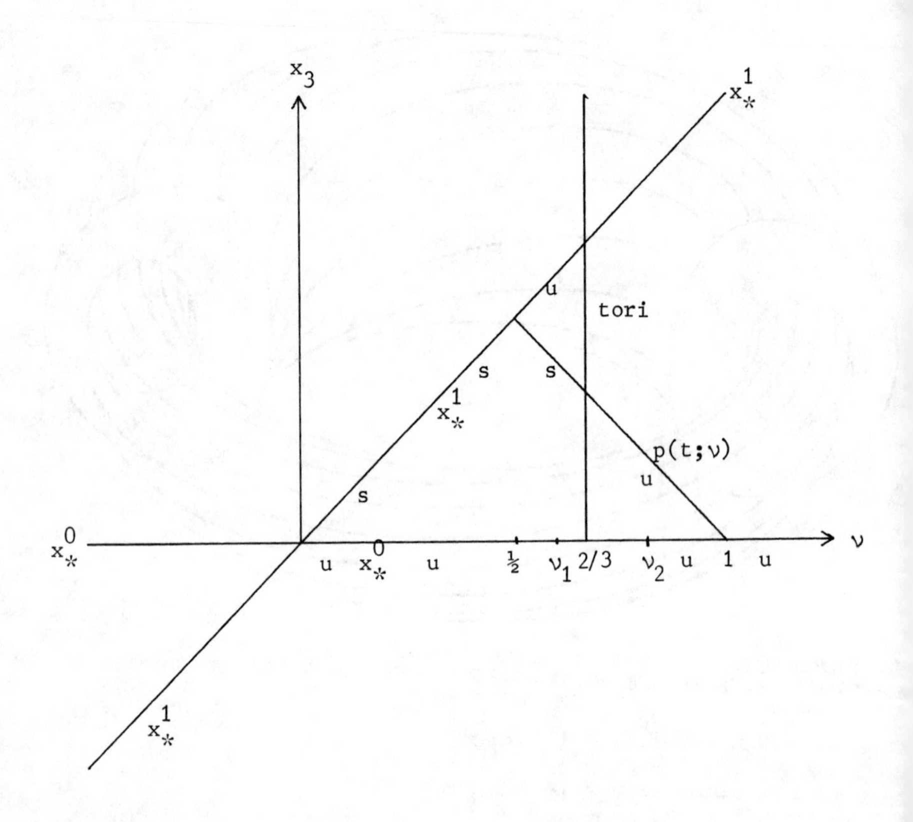

Figure 2.2

A global bifurcation diagram for Langford's system

(s = stable, u = unstable)

3. EXERCISES

1. What will be the computed values of μ_k, τ_k, and β_k for a <u>linear</u> system of ordinary differential equations which satisfies the hypotheses of Hopf's theorem?

2. If two distinct systems of ordinary differential equations

$$\dot{x} = A(\mu)x + f(x;\mu)$$

and

$$\dot{y} = A(\mu)y + g(y;\mu)$$

have a common stationary point

$$x_*(\mu) = y_*(\mu) = 0 ,$$

each satisfies the hypotheses of Hopf's Theorem at $\mu = 0$, and f and g have Taylor expansions about $x_*(\mu)$ (in $x - x_*(\nu)$ and $y - y_*(\nu)$, respectively) that coincide up to and including cubic terms, how will the computed values of μ_2, τ_2, and β_2 for each system compare?

3. Suppose that the system of o.d.e.'s has parameters $\pi_1,\ldots,\pi_m$ in addition to the parameter ν. Describe conditions under which the bifurcation parameters μ_2, τ_2, β_2 will be algebraic functions of $\pi_1,\ldots,\pi_m$.

4. In Table (2.1), the agreement between the columns labelled $x(y = 0)$ and $(4\mu/3)^{1/2}$ is not especially good for larger values of μ. Use the approximation to the periodic solution $p(t;\mu)$ to find the intersection of the periodic solution with the positive x-axis, correct to terms $O(\varepsilon^3)$, and compare your result with the computed value.

5. Consider the predator-prey system [70] modeled by the equations

$$\frac{dN}{d\tau} = aN - eN^2 - bP\phi(N)$$

$$\frac{dP}{d\tau} = -cP + dP\phi(N) \ , \qquad (\dagger)$$

where N is the prey biomass, P is the predator biomass, a, b, c, d and e are positive constants, and $\phi(N)$ represents a functional response of predators to prey. This is an extension of the classical Volterra model $(e = 0, \phi(N) = N)$ which allows competition among prey and which includes a more general interaction term $\phi(N)$, which is assumed positive, increasing, and "smooth".

(a) Substitute $\tau = t/a$, $\tilde{P} = P\,a/b$ in (†) to obtain the system

$$\frac{dN}{dt} = N - \nu N^2 - \tilde{P}\phi(N)$$

$$\frac{d\tilde{P}}{dt} = -\gamma\tilde{P} + \delta\tilde{P}\phi(N) \ ,$$

where $\nu = e/a$, $\gamma = c/a$, $\delta = d/a$. Let $\phi^{-1}(\cdot)$ denote the inverse function for ϕ. There are then stationary points $(N,\tilde{P}) = (0,0)$ and $(N,\tilde{P}) = (N_*,\tilde{P}_*)$, where $N_* = \phi^{-1}(\gamma/\delta)$, $\tilde{P}_* = \frac{\delta}{\gamma}(1 - \nu N_*)N_*$. Assume that $N_* > 0$, $\tilde{P}_* > 0$, and show that the characteristic equation of the Jacobian matrix at the stationary point $(N_*,\tilde{P}_*)$ is

$$\lambda^2 + \lambda[2\nu N_* + \tilde{P}_*\phi'(N_*) - 1] + \gamma\tilde{P}_*\phi'(N_*) = 0 \ .$$

(b) Show that as ν is decreased past

$$\nu_c = \frac{\delta N^*\phi' - \gamma}{N^*(\delta N^*\phi' - 2\gamma)} \ ,$$

a Hopf bifurcation occurs. (Assume that the parameters are such that ν_c is well defined and positive.)

(c) Show that $\mathrm{Re}\, c_1(0)$ is

$$\mathrm{Re}\, c_1(0) = \frac{1}{8}\left[(\nu_c + \frac{P^*\phi''}{2})(\frac{\delta}{\gamma}\phi' + \frac{\phi''}{\phi'}) - \frac{P^*\phi'''}{2}\right] .$$

Suggestion: First examine the formula for $c_1(0)$, and compute only the relevant quantities.

(d) For the functional response

$$\phi(N) = N(1 + \alpha N)^{-1}$$

verify that if $\delta - \alpha\gamma > \gamma\nu$, then

$$\nu_c = \alpha\left[\frac{\delta - \alpha\gamma}{\delta + \alpha\gamma}\right] ,$$

and

$$\mu_2 = -\frac{\alpha^2(\delta - \alpha\gamma)}{2\gamma(\delta + \alpha\gamma)^2}$$

so that stable, small amplitude periodic solutions bifurcate for $\nu < \nu_c$.

5. The Fitzhugh-Nagumo system [57, 88]

$$\dot{x}_1 = \eta + x_1 + x_2 - x_1^3/3$$
$$\dot{x}_2 = \rho(a - x_1 - bx_2) , \qquad \text{(F - N)}$$

where $a,\eta \in \mathbb{R}^1$ and $b,\rho \in (0,1)$, is a prototype for the Hodgkin-Huxley system considered in Chapter 3.

(a) Show that for each $\eta \in \mathbb{R}^1$, (F - N) has a unique stationary point $x_*(\eta)$ for fixed a,b,ρ .

(b) Choose η as the bifurcation parameter. Show that a Hopf bifurcation occurs at $x_*(\eta)$ for two values of η ; namely,

$$\eta_1 = s[\frac{2 + \rho b}{3} - \frac{1}{b}] - \frac{a}{b}$$

and

$$\eta_2 = -\eta_1 - 2a/b ,$$

where $s = (1 - \rho b)^{\frac{1}{2}}$. Show also that

$$x_*(\eta_1) = (-s,(a + s)/b)^T$$

and

$$x_*(\eta_2) = (+s,(a - s)/b)^T .$$

(c) Show that for the bifurcation at η_j (with $\mu = \eta - \eta_j$)

$$\text{Re}(c_1(0)) = \frac{-(1 - 2b + \rho b^2)}{8(1 - \rho b^2)} \qquad (j = 1,2) ,$$

and

$$\mu_2 = - \text{Re}c_1(0)/\alpha'(0) ,$$

where

$$\alpha'(0) = -x_1(\eta_j)x_1'(\eta_j)$$

and

$$x_1'(\eta_j) = -b/(b - 1 - bx_1^2(\eta_j)) .$$

7. Let x_1, x_2, x_3 denote the concentrations of three chemicals c_1, c_2, c_3 , respectively,in a compartment (cell I). Let y_1, y_2, y_3 denote the concentrations of the same three chemicals in the same order in a second compartment (cell II). Suppose that the cells are separated by a membrane through which diffusion takes place

and suppose that $x = (x_1,x_2,x_3)^T$, $y = (y_1,y_2,y_3)^T$ are governed by the differential equations

$$\frac{dx}{dt} = Ax + D(y - x) + C(x)$$

$$\frac{dy}{dt} = Ay + D(x - y) + C(y) ,$$

$$A = \begin{bmatrix} -0.1 & -1 & 0.8 \\ 1 & -0.1 & 0 \\ 0.8 & 0 & -0.1 \end{bmatrix}$$

$$C(x) = \begin{bmatrix} -x_1x_2^2 \\ 0 \\ 0 \end{bmatrix}, \text{ and } D = \begin{bmatrix} .01 & 0 & 0 \\ 0 & .01 & 0 \\ 0 & 0 & \nu/2 \end{bmatrix} .$$

'his example was constructed by L. Howard. Investigate lopf bifurcation for this coupled nonlinear system. In articular, show that there are two critical values of ν

$$\nu_1 = .2314 \quad \text{and} \quad \nu_2 = 1.9953$$

nd find the stability and direction of bifurcation of the eriodic solutions in each case. Hint. Three of the igenvalues are -0.1 and $-0.1 \pm 0.6i$. These are also igenvalues of A . The remaining eigenvalues are also igenvalues of $A - 2D$.

This example is a "life-death" example. If $D = 0$, he origin is stable (globally asymptotically stable in $^3 \times \mathbb{R}^3$) , while if D is as given above there exist stable eriodic solutions. This example is simpler and more xplicit than one constructed for similar purposes by . Smale. Smale determined the global phase portrait for his example [81, pp. 354-367]. But can the global behavior of the nonlinear coupled system of Howard be analyzed and global stability of the periodic orbits be established, at least in the sense that almost all trajectories tend to a periodic orbit if one exists? It would be nice, in our opinion, to construct an example like Howard's except that in the new example, the eigenvalues of A would be all real and negative.

8. Derivation of Recipe-Summary from results of Chapter 1.

Suppose that the system

$$\dot{y} = F(y;\nu) \qquad (y \in \mathbb{R}^n)$$

has $y^*(\nu) = 0$ as an isolated stationary point for all ν in some open interval including $\nu = \nu_c$, and that the Jacobian matrix at $y = 0$ has the real canonical form

$$\frac{\partial F}{\partial y}(0;\nu) = \begin{bmatrix} \alpha & -\omega & \\ \omega & \alpha & 0 \\ & 0 & D \end{bmatrix} .$$

Let $\mu = \nu - \nu_c$, $A(\mu) = \frac{\partial F}{\partial y}(0;\nu_c + \mu)$, $\lambda(\mu) = \alpha + i\omega$, $\lambda_1 = \lambda$, $\lambda_2 = \bar{\lambda}$; and let $\lambda_j(\mu)$, $(j = 3,\ldots,n)$ denote the eigenvalues of D.

(a) Verify that

$$q(\mu) = (1/2,-i/2,0,\ldots,0)^T$$

is an eigenvector of $A(\mu)$ corresponding to λ_1, while

$$q^*(\mu) = (1,-i,0,\ldots,0)^T$$

is an eigenvector of $A^T(\mu)$ corresponding to λ_2.

Also verify that

$$\langle q^*, q\rangle = 1 \ , \quad \langle q^*, \bar{q}\rangle = 0 \ .$$

b) Define

$$z = \langle q^*, y\rangle \ ,$$

$$w = y - zq - \bar{z}\bar{q} \ .$$

Then show that $z = y_1 + iy_2$ and

$$w = \begin{pmatrix} 0 \\ 0 \\ \hat{w} \end{pmatrix} , \quad \text{where} \quad \hat{w} = \begin{pmatrix} y_3 \\ \vdots \\ y_n \end{pmatrix} .$$

(The orthogonality condition (6.8) of Chapter 1 is simply that the first two components of w vanish.)

c) Show that if $y = y(t)$ satisfies

$$\dot{y} = A(\mu)y + f(y;\mu) \ ,$$

where $f(y,\mu) \equiv F(y;\nu_c + \mu) - A(\mu)y$, and if $z(t)$ and $\hat{w}(t)$ are defined in terms of y as in (b), then

$$\dot{z} = \lambda z + G(z,\bar{z},\hat{w};\mu) \ ,$$

$$\dot{\hat{w}} = D\hat{w} + \hat{H}(z,\bar{z},\hat{w};\mu) \ ,$$

where

$$G(z,\bar{z},\hat{w};\mu) = \tfrac{1}{2}[F^1 + iF^2 - \lambda(y_1 + iy_2)] \ ,$$

$$\hat{H}(z,\bar{z},\hat{w};\mu) = \hat{F} - D\hat{w} \ , \text{ and}$$

$$\hat{F}(y;\mu) = (F^3,\ldots,F^n)^T \ .$$

d) We see from part (b) that the variables y_1, y_2, z and $\bar{z}$ are related by

$$y_1 = (z + \bar{z})/2 \ , \quad y_2 = (z - \bar{z})/2i \ .$$

Show that, symbolically,

$$\frac{\partial}{\partial z} = \frac{1}{2}\left(\frac{\partial}{\partial y_1} - i\,\frac{\partial}{\partial y_2}\right), \quad \frac{\partial}{\partial \bar{z}} = \frac{1}{2}\left(\frac{\partial}{\partial y_1} + i\,\frac{\partial}{\partial y_2}\right).$$

That is, if (i) $\varphi(y_1,y_2)$ is a function of <u>complex</u> variables y_1 and y_2, (ii) $\psi(\xi,\eta)$ is defined in terms of φ by means of $\psi(\xi,\eta) =$ $\varphi((\xi + \eta)/2,(\xi - \eta)/2i)$, (iii) each of $\partial\psi(z,\bar{z})/\partial\xi$, $\partial\psi(z,\bar{z})/\partial\eta$, $\partial\varphi(y_1,y_2)/\partial y_1$, and $\partial\varphi(y_1,y_2)/\partial y_2$ is meaningful for each of y_1 and y_2 real, and (iv) $z = y_1 + iy_2$ and $\bar{z} = y_1 - iy_2$, then

$$\frac{\partial\psi}{\partial\xi}(z,\bar{z}) = \frac{1}{2}\left(\frac{\partial\varphi}{\partial y_1}(y_1,y_2) - i\,\frac{\partial\varphi}{\partial y_2}(y_1,y_2)\right),$$

$$\frac{\partial\psi}{\partial\eta}(z,\bar{z}) = \frac{1}{2}\left(\frac{\partial\varphi}{\partial y_1}(y_1,y_2) + i\,\frac{\partial\varphi}{\partial y_2}(y_1,y_2)\right).$$

<u>Remark</u>. The point to be noticed is that if $\varphi(y_1,y_2)$ is defined only for <u>real</u> y_1,y_2, the right-hand sides of the two equations above serve to <u>define</u> the operators $\partial/\partial z$ and $\partial/\partial\bar{z}$.

(e) Show that, symbolically,

$$4\,\frac{\partial^2}{\partial z^2} = \frac{\partial^2}{\partial y_1^2} - \frac{\partial^2}{\partial y_2^2} - 2i\,\frac{\partial^2}{\partial y_1 \partial y_2},$$

$$4\,\frac{\partial^2}{\partial \bar{z}^2} = \frac{\partial^2}{\partial y_1^2} - \frac{\partial^2}{\partial y_2^2} + 2i\,\frac{\partial^2}{\partial y_1 \partial y_2},$$

$$4\,\frac{\partial^2}{\partial z \partial \bar{z}} = \frac{\partial^2}{\partial y_1^2} + \frac{\partial^2}{\partial y_2^2}.$$

Then obtain formulae for the second "partial

derivatives"

$$g_{ij} \equiv \frac{\partial^2 G}{\partial z^i \partial \bar{z}^j}(0,0,0;\mu) \quad (i + j = 2) \ ,$$

$$\hat{h}_{ij} \equiv \frac{\partial^2 \hat{H}}{\partial z^i \partial \bar{z}^j}(0,0,0;\mu) \quad (i + j = 2)$$

in terms of the second partials of the functions F^k $(k = 1,\ldots,n)$. (G and $\hat{H}$ were defined in part (c) above, and each of the $\hat{h}_{ij}$'s are $n - 2$ dimensional vectors.)

(f) Since the real, n-dimensional, vector-valued function

$$w = w(z,\bar{z};\mu) \ ,$$

representing the center manifold for the system in the form (6.6) of Chapter 1 satisfies $\langle q^*,w\rangle = 0$, one may write

$$w(z,\bar{z};\mu) = \begin{pmatrix} 0 \\ 0 \\ \hat{w}(z,\bar{z};\mu) \end{pmatrix} ,$$

where $\hat{w}(z,\bar{z};\mu)$ is real, $(n - 2)$-dimensional, and represents the center manifold for the system in (c) above.

Use the expressions for the expansion coefficients of $w(z,\bar{z};\mu)$ given in Chapter 1, Section 6 to show that

$$w_{ij}(\mu) = \begin{pmatrix} 0 \\ 0 \\ \hat{w}_{ij}(\mu) \end{pmatrix} ,$$

where, for $i + j = 2$,

$$\hat{w}_{ij}(\mu) = [(\lambda i + \bar{\lambda} j)I - D]^{-1}\hat{h}_{ij}(\mu) \ ,$$

where

$$\hat{h}_{ij}(\mu) = \partial^2\hat{h}(0,0;\mu)/\partial z^i \partial \bar{z}^j$$

and

$$h(z,z;\mu) = \begin{pmatrix} 0 \\ 0 \\ \hat{h}(z,\bar{z};\mu) \end{pmatrix} .$$

(g) Let

$$g(z,\bar{z};\mu) \equiv G(z,\bar{z},\hat{w}(z,\bar{z};\mu);\mu) .$$

Show that

$$g(z,\bar{z};\mu) = G(z,\bar{z},\mathfrak{J}_2\hat{w}(z,\bar{z};\mu);\mu) + O(|z|^4) ,$$

where

$$\mathfrak{J}_2\hat{w}(z,\bar{z};\mu) \equiv \hat{w}_{20}z^2/2 + \hat{w}_{11}z\bar{z} + \hat{w}_{02}\bar{z}^2/2$$

denotes the quadratic terms in the Taylor expansion of $\hat{w}(z,\bar{z};\mu)$.

(h) Show that, symbolically

$$8\frac{\partial^3}{\partial z^2 \partial \bar{z}} = \frac{\partial^3}{\partial y_1^3} + \frac{\partial^3}{\partial y_1 \partial y_2^2} - i\left(\frac{\partial^3}{\partial y_1^2 \partial y_2} + \frac{\partial^3}{\partial y_2^3}\right) ;$$

then obtain the formula for

$$G_{21} = \frac{\partial^3 G}{\partial z^2 \partial \bar{z}}(0,0,0;\mu)$$

in terms of third partials of F^1 and F^2.

(i) Obtain formulae for the elements of the (n - 2)-dimensional vectors

$$G_{110}(\mu) \equiv \frac{\partial^2 G}{\partial \hat{w} \partial z}(0,0,0;\mu) ,$$

$$G_{101}(\mu) \equiv \frac{\partial^2 G}{\partial \hat{w} \partial \bar{z}}(0,0,0;\mu)$$

in terms of second partials of F^1 and F^2.

(j) Define

$$g_{21}(\mu) \equiv \frac{\partial^2}{\partial z^2 \partial \bar{z}} g(0,0;\mu) .$$

By part (g), $g_{21}(\mu)/2$ may be computed as the coefficient of $z^2\bar{z}$ in the Taylor expansion of

$$G(z,\bar{z},\mathfrak{J}_2\hat{w}(z,\bar{z};\mu);\mu) .$$

Show that since

$$G(z,\bar{z},\hat{w};\mu) = G(z,\bar{z},0;\mu) + \sum_{k=1}^{n-2} \frac{\partial G}{\partial \hat{w}_k}(z,\bar{z},0;\mu)\hat{w}_k$$

$$+ O(|\hat{w}|^2)$$

and $\hat{w}(z,\bar{z};\mu) = O(|z|^2)$,

$$g_{21}(\mu) = G_{21}(\mu) + 2 \sum_{k=1}^{n-2} \frac{\partial^2 G}{\partial \hat{w}_k \partial z}(0,0,0;\mu)\hat{w}_{11}^k(\mu)$$

$$+ \sum_{k=1}^{n-2} \frac{\partial^2 G}{\partial \hat{w}_k \partial \bar{z}}(0,0,0;\mu)\hat{w}_{20}^k(\mu) .$$

Thus

$$g_{21} = G_{21} + \sum_{k=1}^{n-2} [2G_{110}^k \hat{w}_{11}^k + G_{101}^k \hat{w}_{20}^k] .$$

(k) Check your answers in parts (e), (f), (h) and (i) by

setting $\mu = 0$ and comparing with the formulae in steps 6 and 7 of the Recipe-Summary.

(ℓ) Convince yourself that the same formulae for μ_2, τ_2 and β_2 would result if the entire manipulation had been performed without the aid of the symbolic operators $\partial/\partial z$, $\partial/\partial \bar{z}$.

CHAPTER 3. NUMERICAL EVALUATION OF HOPF BIFURCATION FORMULAE

1. INTRODUCTION

For problems beyond a certain complexity, it becomes impossible to evaluate bifurcation formulae analytically, i.e. in terms of closed form, analytical expressions. In Chapter 2, the examples presented were of orders 2 and 3 because the amount of algebraic manipulation involved increases rapidly with N, the order of the system. The present chapter considers the numerical evaluation of formulae for the Hopf bifurcation for systems of autonomous ordinary differential equations, of the form

$$\dot{x} = f(x;\nu), \quad x \in \mathbb{R}^N . \tag{1.1}$$

We describe the required calculations, give FORTRAN codes to solve the general problem, and present sample applications, including applications to partial differential systems in one space dimension. Applications to delay and integrodifferential systems are outlined in the Exercises for Chapter 4.

To gain perspective, it is useful to compare the present technique with direct numerical integration for the periodic solutions as, for example, in [44]. The result of numerical integration is a large set of numbers representing the solution for a fixed value of the parameter ν. The integration is repeated to obtain a one parameter family of solutions. In contrast, the result of evaluating bifurcation formulae is a small set of numbers, the coefficients in the approximations

$$x(t,\nu) = x_*(\nu_c) + \left[\frac{\nu - \nu_c}{\mu_2}\right]^{\frac{1}{2}} \mathrm{Re}(e^{2\pi it/T} v_1) + O(\nu - \nu_c) \quad (1.2)$$

$$T(\nu) = \frac{2\pi}{\omega_0}[1 + \tau_2\left[\frac{\nu - \nu_c}{\mu_2}\right] + O(\nu - \nu_c)^2] , \quad (1.3)$$

$$\beta(\nu) = \beta_2\left[\frac{\nu - \nu_c}{\mu_2}\right] + O(\nu - \nu_c)^2 , \quad (1.4)$$

valid for $\mu_2 \neq 0$. Here $T(\nu)$ is the period of the periodic solutions and $\beta(\nu)$ is the characteristic exponent that determines their orbital stability. If $\beta(\nu) < 0$, the periodic solutions are orbitally asymptotically stable with asymptotic phase. Once ν_c, μ_2, v_1, ω_0, τ_2 and β_2 have been calculated, a one parameter family of solutions is known to the degree of approximation shown. The condensation of information down to this set of numbers is quite dramatic: it must be kept in mind, however, that the information is local. For any fixed value of ν near ν_c, careful numerical integration will provide more accuracy than the given approximations. Also, for larger values of $|\nu - \nu_c|$, numerical integration is often the only means of obtaining quantitative information about the periodic solutions. However, numerical integration for the family of periodic solutions can become expensive, since one is solving a one parameter family of nonlinear boundary value problems with periodic boundary conditions. The computational effort involved in the numerical evaluation of bifurcation formulae is slightly more than in linear stability analysis of the stationary solutions. A logical way to proceed is to use the techniques of the present Chapter to uncover the "interesting" combinations of parameters in the system for which numerical integrations will then be performed. For example, in designing a control system which must operate "near the edge" of instability, one may use the present techniques to define the range of physical parameters such that should a loss of stability by the mechanism of Hopf bifurcation occur, the periodic solutions which arise are themselves stable rather than unstable. Simulations may then be run for parameters in this range.

The codes we present for the numerical evaluation of Hopf bifurcation formulae require that the user supply a main program to call the routine BIFOR2 and a subroutine to evaluate both the vector function $f(x,\nu)$ defining the system of o.d.e.'s and the Jacobian matrix of first partial derivatives $\partial f/\partial x(x;\nu)$. Also required are estimates of the critical value of the bifurcation parameter and of the stationary point. The array storage requirements are $2N^2 + 15N$, and the amount of computational effort grows roughly as $O(N^3)$. BIFOR2 satisfies the FORTRAN standard ANSI X3.9 (1966) .

Our intent in the present Chapter is to describe how BIFOR2 works and to present sample applications, so that anyone with moderate FORTRAN ability can run Hopf bifurcation computations. A description of the parameters of BIFOR2 is given in Appendix E , and a sample program written to call BIFOR2 is given in Appendix F. The complete listing of BIFOR2 is on the microfiche. Also on microfiche is the listing of a driver program, which produces most of the numerical results in the present Chapter, and in addition verifies most of the hand calculations of Chapters 2, 4 and 5. In the following, when we mention a subroutine or function subprogram by name, the actual code may be found by referring first to Appendix G, then to the microfiche.

2. LOCATION OF THE CRITICAL VALUE OF THE BIFURCATION PARAMETER AND THE CORRESPONDING EQUILIBRIUM SOLUTION

BIFOR2 offers the choice (MTH = 1 or 2) of two different methods of locating ν_c . The first method was designed for reliability and offers the option of displaying all the eigenvalues of the Jacobian matrix at each iterate. The second method does not provide this option and is generally faster but slightly less reliable. Both methods are performed by subroutine ANUCRT.

Method 1 for location of ν_c.

The critical value of ν is located by solving the algebraic equation

$$\alpha(\nu) = 0 \tag{2.1}$$

by means of the secant method. Here $\alpha(\nu) = \text{Re}\,\lambda_1(\nu)$, where $\lambda_1(\nu)$ denotes the eigenvalue of the Jacobian matrix

$$A(\nu) = \frac{\partial f}{\partial x}(x_*(\nu);\nu) \tag{2.2}$$

given by

$$\text{Re}\,\lambda_1 = \max\{\text{Re}\,\lambda | \lambda \in \sigma(A)\},$$

$$\text{Im}\,\lambda_1 = \max\{\text{Im}\,\lambda | \lambda \in \sigma(A),\ \text{Re}\,\lambda = \text{Re}\,\lambda_1\}, \tag{2.3}$$

and $\sigma(A)$ denotes the set of eigenvalues of A. At each iterate ν_k in the location of ν_c by the secant method, ANUCRT calls subroutine EVALS which in turn calls subroutine NWTN (Newton's method) to compute $x_*(\nu_k)$ and then calls subroutine EIGR (the QR algorithm) to compute the eigenvalues of $A(\nu_k)$.

The secant iteration.

The secant iteration is given by

$$\nu_{k+1} = \nu_k - \gamma_k(\nu_k - \nu_{k-1}) \quad (k = 1,2,\ldots), \tag{2.4}$$

where

$$\gamma_k = \alpha(\nu_k)/(\alpha(\nu_k) - \alpha(\nu_{k-1})). \tag{2.5}$$

Before describing the computation of $\alpha(\nu_k)$, we shall discuss how the secant iteration is started and stopped.

To start the iteration, a pair of values ν_0 and ν_1 are eeded. For ν_1, the value taken is the user-supplied estimate of ν_c. Then ν_0 is taken as $\nu_0 = \nu_1 + \Delta\nu$, where $\Delta\nu$ is given by

$$\Delta\nu = u^{\frac{1}{2}}\nu_{ref} , \tag{2.6}$$

n which u is an estimate of the relative machine precision and $\nu_{ref} = ref(\nu_1)$, where $ref(\cdot)$ is the function

$$ref(x) = \begin{cases} |x| & \text{if } |x| \geq \epsilon_{ref} , \\ 1 & \text{if } |x| < \epsilon_{ref} . \end{cases} \tag{2.7}$$

The construction of ν_{ref} from ν_1 incorporates the idea hat the increment $\Delta\nu$ should be taken relative to the magnitude f ν_1 because $|\nu_1|$ is normally expected to represent a scale or the variable ν. When $\nu_1 = 0$, however, $|\nu_1|$ provides no uch information about an appropriate scale for ν. In this ase, the function $ref(\cdot)$ assigns the value $\nu_{ref} = 1.0$, a easonable choice in the absence of better information. The umber ϵ_{ref} is the criterion used to decide whether or not ν_1 as intended to be 0, and is defined within function REF.

The relative machine precision is the smallest number u uch that the machine employed distinguishes between $(1.0 + u)$ nd 1.0, and the power $u^{\frac{1}{2}}$ in (2.6) arises from the idea that he quotient $(\alpha(\nu_0) - \alpha(\nu_1))/\Delta\nu$ is a one-sided difference pproximation for $\alpha'(\nu_1)$. For elaboration of this idea, see ppendix D.

The iteration for ν_c is normally stopped when either

$$\begin{aligned} &1) \quad |\alpha(\nu_k)| < \epsilon , \text{ or} \\ &2) \quad |(\nu_k - \nu_{k-1})/\nu_{ref}| < 10^{-N_{sig}} . \end{aligned} \tag{2.8}$$

Here ϵ is "small", and N_{sig} is the number of decimal digits of relative precision desired. The values for ϵ and N_{sig} are defined within subroutine BIFOR2. An abnormal termination and error return from ANUCRT occurs when $k > \text{ITMAX}$. The value for ITMAX is defined within BIFOR2.

<u>Solution for</u> $x_*(\nu_k)$.

At each iterate $\nu = \nu_k$, the stationary point $x_*(\nu_k)$ is computed by Newton's method. (The Hopf hypotheses, together with continuity of $\partial f/\partial x(x,\nu)$, guarantee the invertibility of the Jacobian matrix for all (x,ν) sufficiently close to $(x_*(\nu_c),\nu_c)$.) The user-supplied estimate of $x_*(\nu_c)$ is the initial point in the Newton iteration for $x(\nu_1)$, and the result x_*^1 of this iteration is used as the initial approximation in the Newton iteration for $x(\nu_0)$. Let x_*^0 denote the result of the iteration for $x_*(\nu_0)$. For $k = 1,2,\dots$, the point

$$x_0^{k+1} = x_*^k - \gamma_k(x_*^k - x_*^{k-1}) \tag{2.9}$$

is used as the initial approximation to $x_*(\nu_{k+1})$, where γ_k is from (2.5), and x_*^k denotes the result of the Newton iteration for $x_*(\nu_k)$.

For fixed $k \geq 0$, let x_j^k $(j = 0,1,2,\dots)$ denote the sequence generated in the Newton iteration for $x_*(\nu_k)$; and let $(x_j^k)_i$ $(1 \leq i \leq N)$ denote the components of the vector x_j^k . The iteration for $x_*(\nu_k)$ is then

$$x_{j+1}^k = x_j^k - [\frac{\partial f}{\partial x}(x_j^k;\ \nu_k)]^{-1} f(x_j^k;\ \nu_k) \qquad (j = 0,1,\dots) . \tag{2.10}$$

The iteration is normally stopped at the first j such that either

$$1)\quad \|f(x_j^k;\ \nu_k)\| < \varepsilon\ ,\ \text{or}$$

$$2)\quad \sum_{i=1}^{N} ((x_j^k - x_{j-1}^k)_i / x_{ref}^i)^2 < 10^{-2N_{sig}}\ , \tag{2.11}$$

where ε and N_{sig} are as above and x_{ref} is an N-vector, the components of x_{ref} being scales for the corresponding components of x. The components of x_{ref} are computed within BIFOR2 from the corresponding components of the user-supplied estimate for $x_*(\nu_c)$, using the function ref(·) described above. The current iterate x_j^k when the iteration is stopped is taken as the approximation x_*^k to the solution $x_*(\nu_k)$.

The Newton iteration itself is performed within subroutine NWTN. LINPACK [29] routines SGEFA and SGESL are used to solve the linear systems

$$\frac{\partial f}{\partial x}(x_j^k;\ \nu_k)s_k = f(x_j^k;\ \nu_k)$$

indicated by (2.10), and then x_{j+1}^k is defined by

$$x_{j+1}^k = x_j^k - s_k\ .$$

The Newton iteration terminates abnormally and an error return from NWTN occurs when $j > \text{ITMAX}$, where ITMAX is as above.

<u>Computation of</u> $\lambda_1(\nu_k)$.

After $x_*(\nu_k)$ is found as above, the Jacobian matrix $J_k = \partial f/\partial x(x_*^k;\ \nu_k)$ is evaluated, and then the double-step QR algorithm [29, 110] (EISPACK [100] subroutine HQR, called by subroutine EIGR) is used to compute all the eigenvalues of this matrix. The eigenvalue $\lambda_1(\nu_k)$ is then selected according (2.3).

In Method 1, the computation of $\lambda_1(\nu_k)$ is organized by subroutine EVALS, which calls NWTN to find $x_*(\nu_k)$, then EIGR to find λ_1 .

<u>Method 2 for location of</u> ν_c .

This method differs from the method described above only in the technique used to solve the eigenvalue problems. The secant method is again performed by subroutine ANUCRT. Just as above, the QR algorithm (subroutine EIGR, called by EVALS) is used to evaluate $\lambda_1(\nu_1)$. Then, however, inverse iteration [27, 91] rather than the QR algorithm is used to evaluate $\lambda_1(\nu_0)$, and in addition $v_1(\nu_0)$, the corresponding eigenvector. Since ν_0 is close to ν_1 , $\lambda_1(\nu_0)$ is in general close to $\lambda_1(\nu_1)$; and the inverse iteration for $\lambda_1(\nu_0)$, based upon the approximation $\lambda_1(\nu_0) \approx \lambda_1(\nu_1)$, tends to converge rapidly even though an arbitrary initial guess is used for $v_1(\nu_0)$. The inverse iteration is performed by subroutine INITER which is called by EVAL1. Inverse iteration is also used for the evaluation of $\lambda_1(\nu_{k+1})$ for $k \geq 1$, and the iteration is based upon the (extrapolated) approximation

$$\lambda_1(\nu_{k+1}) \approx \lambda_1(\nu_k) - \gamma_k(\lambda_1(\nu_k) - \lambda_1(\nu_{k-1})) \quad (k = 1,2,\ldots), \tag{2.12}$$

where γ_k is as in (2.5). The extrapolation

$$v_1(\nu_{k+1}) \approx v_1(\nu_k) - \gamma_k(v_1(\nu_k) - v_1(\nu_{k-1})) \tag{2.13}$$

provides the initial approximation for the eigenvector $v_1(\nu_{k+1})$, $k \geq 3$. For $k = 1$ and 2 , the same formula is employed except that $v_1(\nu_0)$ is used instead of $v_1(\nu_1)$, which was not calculated.

The difference in execution times between Methods 1 and 2 will depend both upon the machine and upon the problem. In the Hodgkin-Huxley system the expense in evaluating the Jacobian matrix dominates the expense of solving the algebraic eigenvalue problem, so use of the second method rather than the first makes little difference in execution time. For the panel flutter problem, we have observed Method 2 to be twice as fast as Method 1; and so, for this problem, a large fraction of the computational effort is evidently spent on the algebraic eigenvalue problem.

At one stage in the development of BIFOR2, we investigated an algorithm for the location of ν_c closely related to one recently proposed by Kubiček [73]. See Appendix E for a brief description of this algorithm, and the reasons we have not included it in BIFOR2.

3. EVALUATION OF THE COEFFICIENT $c_1(0)$ OF THE POINCARÉ NORMAL FORM

Once ν_c, $x_*(\nu_c)$, and $\omega_0 = \omega(\nu_c)$ have been determined, BIFOR2 calls subroutine C1PNF to evaluate the coefficient $c_1(0)$ of the Poincaré normal form.

The stages in evaluating $c_1(0)$ are as follows.

1) Find the right and left eigenvectors v_1 and u_1 of the Jacobian matrix $A = \partial f/\partial x(x_*(\nu_c); \nu_c)$ which correspond to the eigenvalue $\lambda_1(\nu_c) = i\omega_0$.

2) Normalize v_1 so that the first nonvanishing element is identically 1, and normalize u_1 relative to v_1 so that $u_1^T v_1 = 1$.

3) Perform numerical differencing to approximate the second partial derivatives

$$\begin{aligned} f_{20} &= (\partial^2/\partial z^2) f(x_* + \mathrm{Re}(v_1 z); \nu_c)|_{z=0} \\ f_{11} &= (\partial^2/\partial z \partial \bar{z}) f(x_* + \mathrm{Re}(v_1 z); \nu_c)|_{z=0} . \end{aligned} \tag{3.1}$$

4) Calculate

$$\begin{aligned} g_{20} &= 2u_1^T f_{20}, \ g_{11} = 2u_1^T f_{11}, \ g_{02} = 2u_1^T \bar{f}_{20} \\ h_{20} &= P_\perp f_{20}, \ h_{11} = P_\perp f_{11} , \end{aligned} \tag{3.2}$$

where

$$P_{\perp} = I - 2\,\mathrm{Re}(v_1 u_1^T) .$$

5) Solve the linear systems

$$A\,w_{11} = -h_{11} ,$$

$$(A - 2i\omega_0 I)\,w_{20} = -h_{20} , \tag{3.3}$$

for the coefficient vectors w_{11}, w_{20} of the quadratic terms in expansion of the slice $\nu = \nu_c$ of the center manifold.

6) Perform numerical differencing to evaluate the third partial derivative

$$G_{21} = (\partial^3/\partial z^2\partial\bar{z})[2u_1^T f(x_* + \mathrm{Re}(v_1 z + w_{20}z^2 + w_{11}z\bar{z});\ \nu_c)]|_{z=0} . \tag{3.4}$$

7) Evaluate

$$c_1(0) = \frac{i}{2\omega_0}[g_{20}g_{11} - 2|g_{11}|^2 - \frac{1}{3}|g_{02}|^2] + \frac{G_{21}}{2} . \tag{3.5}$$

<u>Remark</u>. The derivation of this particular algorithm from the results of Chapter 1 is assigned as Exercise 1 at the end of the present Chapter.

3.1 Right and left eigenvectors of A

Inverse iteration is used to compute the right eigenvector v_1 of the Jacobian matrix $A = \partial f/\partial x(x_*(\nu_c);\ \nu_c)$, corresponding to the eigenvalue $\lambda_1(\nu_c) = i\omega_0$. An arbitrary initial guess is made for the eigenvector, then ClPNF calls subroutine INITER to perform inverse iteration. In INITER, the initial estimate of the eigenvalue is slightly degraded before being used. (In our original version of INITER, the estimate of the eigenvalue was not degraded. When applied to the mass-spring-belt problem, however, this version failed: we leave it as an exercise for the reader to explain why.) The members of the sequence of vectors generated by the inverse iteration are normalized by subroutine ENRML to have Euclidean norm 1, and so that the component of largest complex magnitude which has the lowest index, is real and positive. The iteration is stopped the first time that the Euclidean norm of the difference between successive approximations to the eigenvector is less than $10^{-N_{sig}}$.

Inverse iteration is similarly used to compute the left eigenvector u_1. Just one LU factorization is performed in computing both v_1 and u_1. LINPACK routines CGEFA and CGESL are used by INITER. CGEFA performs the LU factorization; then CGESL solves for the successive iterates.

3.2 Normalization of v_1 and u_1

The approximation to the periodic solution is of the form

$$x(t; \nu) = x_*(\nu) + \mathrm{Re}(v_1 z + w_{20} z^2 + w_{11} z\bar{z}) + O(|z|^3), \quad (3.6)$$

where $z = \varepsilon \exp(2\pi it/T(\nu)) + O(\varepsilon^2)$. If v_1 is normalized so the first component is 1, 2ε represents (to within $O(\varepsilon^2)$) the peak to peak (max minus min) amplitude of the first component of the periodic solution. It may not be possible to normalize v_1 in this fashion, however. The scheme used (subroutine BFNRML), therefore normalizes v_1 so that the first nonvanishing component is identically 1. (Here, "normalize" means only multiplication by a nonzero scalar.)

Under the Hopf hypotheses, $\lambda_1(\nu_c) = i\omega_0$ is a simple eigenvalue of the Jacobian matrix $A(\nu_c)$. Thus the right and left eigenvectors v_1 and u_1 are uniquely defined, up to multiplication by nonzero scalars, and $u_1^T v_1 \neq 0$. For subsequent purposes, the desired normalization of u_1 is such that $u_1^T v_1 = 1$. This normalization is performed by subroutine RLNRML.

3.3. Numerical differencing for second partial derivatives

The second partial derivatives

$$f_{20} = (\partial^2/\partial z^2) f(x_* + \mathrm{Re}(v_1 z); \nu_c)|_{z=0} ,$$

$$f_{11} = (\partial^2/\partial z \partial \bar{z}) f(x_* + \mathrm{Re}(v_1 z); \nu_c)|_{z=0}$$

are computed by differencing the first partial derivatives as evaluated by the user-supplied subroutine. Let $z = y_1 + iy_2$. Then, symbolically,

$$\partial/\partial z = (\partial/\partial y_1 - i\partial/\partial y_2)/2, \ \partial/\partial \bar{z} = (\partial/\partial y_1 + i\partial/\partial y_2)/2 ,$$

$$\partial^2/\partial z^2 = (\partial^2/\partial y_1^2 - \partial^2/\partial y_2^2 - 2i\ \partial^2/\partial y_1\, \partial y_2)/4$$

$$\partial^2/\partial z \partial \bar{z} = (\partial^2/\partial y_1^2 + \partial^2/\partial y_2^2)/4 ;$$

and the derivatives to evaluate are,

$$(\partial^2/\partial y_j\, \partial y_k) f(x_* + v^r y_1 - v^i y_2)|_{y_1=y_2=0} \qquad (j,k = 1,2) ,$$

where

$$v^r = \mathrm{Re}\ (v_1), \quad v^i = \mathrm{Im}\ (v_1) .$$

Now

$$\frac{\partial}{\partial y_1} f(x + v^r y_1 - v^i y_2) = [\frac{\partial f}{\partial x} (x_* + v^r y_1 - v^i y_2)]\ v^r ,$$

$$\frac{\partial}{\partial y_2} f(x + v^r y_1 - v^i y_2) = -[\frac{\partial f}{\partial x} (x_* + v^r y_1 - v^i y_2)]\ v^i ,$$

where $\partial f/\partial x$ denotes the Jacobian matrix. The matrices

$$A_{\pm} = \partial f/\partial x\ (x_* \pm \Delta y\ v^r)$$

are formed, and the approximations

$$\partial^2 f/\partial y_1^2 \ (x_*) \approx [(A_+ - A_-)v^r]/(2\Delta y) \ ,$$

$$\partial^2 f/\partial y_1 \partial y_2 (x_*) \approx -[(A_+ - A_-)v^i] \ (2\Delta y)$$

re employed. Then, the matrices

$$A_{\pm} = \partial f/\partial x (x_* \pm \Delta y(-v^i))$$

re formed, and the approximation

$$\partial^2 f/\partial y_2^2 (x_*) \approx -[(A_+ - A_-)v^i]/(2\Delta y)$$

s employed. The computation of f_{20} and f_{11} as just escribed is performed within subroutine DIF2.

The increment used is

$$\Delta y = u^{1/3} \text{ ref } (\|x_*\|/\|v_1\|) \ .$$

he factor $u^{1/3}$ arises because the truncation error in the ifference quotients is $O((\Delta y)^2)$, and the roundoff error is $(u/\Delta y)$: see Appendix D. The ratio $\|x_*\|/\|v_1\|$ gives an appropriate scale for Δy , provided $\|x_*\| \neq 0$; if $\|x_*\| = 0$, the cale 1 is assigned by the function $\text{ref}(\cdot)$.

.4. Evaluation of g_{20}, g_{11}, g_{02}, h_{20} and h_{11}

The functions $g(z, \bar{z}, w)$ and $h(z, \bar{z}, w)$ are defined in erms of f by means of

$$g(z, \bar{z}, w) = 2u_1^T f(x_* + (v_1 z + \bar{v}_1 \bar{z})/2 + w) \ ,$$

$$h(z, \bar{z}, w) = P_\perp f(x_* + (v_1 z + \bar{v}_1 \bar{z})/2 + w) \ ,$$

here $P_\perp$ is the real matrix

$$P_\perp = I - 2 \text{ Re } (v_1 u_1^T) \ .$$

he partial derivatives

$$g_{20} = \partial^2 g/\partial z^2, \quad g_{11} = \partial^2 g/\partial z \partial \bar{z}, \quad g_{02} = \partial^2 g/\partial \bar{z}^2$$

$$h_{20} = \partial^2 h/\partial z^2, \quad h_{11} = \partial^2 h/\partial z \partial \bar{z}$$

at $z = w = 0$ are all computed in terms of

$$f_{20} = \partial^2 f/\partial z^2, \quad f_{11} = \partial^2 /\partial z \partial \bar{z}$$

at $z = w = 0$, which were approximated above. Specifically,

$$g_{20} = 2u_1^T f_{20}, \quad g_{11} = 2u_1^T f_{11}, \quad g_{02} = 2u_1^T \bar{f}_{20},$$

$$h_{20} = P_\perp f_{20} \quad \text{and} \quad h_{11} = P_\perp f_{11}.$$

This straightforward computation is performed by subroutine PRJCT2.

3.5. Solution for the coefficient vectors w_{11} and w_{20} in the expansion of the slice $\nu = \nu_c$ of the center manifold.

The vectors w_{11} and w_{20} are solutions of the N-dimensional linear systems

$$Aw_{11} = -h_{11},$$

$$(A - 2i\omega_0 I) w_{20} = -h_{20},$$

where A is the Jacobian matrix $A = \partial f/\partial x \, (x_*(\nu_c); \nu_c)$. The system for w_{11} has a real coefficient matrix, while the system for w_{20} involves a complex matrix. The Hopf hypotheses imply that both coefficient matrices are invertible, and so solution for w_{11} and w_{20} is a straightforward task. Subroutine CMAN2 sets up the linear systems and calls LINPACK subroutines SGEFA, SGESL, CGEFA and CGESL to solve the systems.

3.6. Numerical Differencing for G_{21}

Let

$$G(z,\bar{z}) = g(z,\bar{z},w(z,\bar{z})),$$

where the real N-dimensional vector-valued function $w(z,\bar{z})$ represents the slice $\nu = \nu_c$ of the center manifold. The coefficient $c_1(0)$ in the Poincaré normal form involves the 3rd partial derivative

$$G_{21} = \partial^3 G/\partial z^2 \partial \bar{z}$$

at $z = 0$, $w = 0$. Since $\partial g/\partial w = 0$ at the origin, G_{21} may be computed as

$$G_{21} = \frac{\partial^3}{\partial z^2 \partial \bar{z}} g(z,\bar{z},w_2(z,\bar{z}))\Big|_{z=0} ,$$

where

$$w_2(z,\bar{z}) = w_{20}z^2/2 + w_{11}z\bar{z} + w_{02}\bar{z}^2/2$$

is the quadratic approximation to $w(z,\bar{z})$. Approximations to the coefficient vectors w_{20}, w_{11} were obtained above; and since w is a real vector-valued function, $w_{02} = \bar{w}_{20}$.

Let $z = y_1 + iy_2$. Then, symbolically, $\partial^2/\partial z \partial \bar{z} = (\partial^2/\partial y_1^2 + \partial^2/\partial y_2^2)/4 = \Delta/4$, where Δ denotes the Laplacian in y_1 and y_2. Now

$$(\partial/\partial z)g(z,\bar{z},w_2(z,\bar{z})) = 2u_1^T \frac{\partial f}{\partial x} (x_* + (v_1 z + \bar{v}_1\bar{z})/2 + w_2(z,\bar{z})) \times$$
$$\times (v_1/2 + w_{20}\bar{z} + w_{11}z) - i\omega_0 .$$

Subroutine DIF3 approximates G_{21} by applying a finite difference operator, the 9-point Laplacian $\Delta_9(h)$ in the variables $y_1 = \mathrm{Re}(z)$ and $y_2 = \mathrm{Im}(z)$, to the function $(\partial/\partial z)g(z,\bar{z},w_2(z,\bar{z}))$, which is evaluated as above. The 9-point Laplacian is constructed as the Richardson extrapolation

$$\Delta_9(h) = (4\Delta_5(h) - \Delta_5(2h))/3$$

of the customary 5-point Laplacian

$$\Delta_5(h)\psi(0,0) = (\psi(h,0) + \psi(-h,0) + \psi(0,h) + \psi(0,-h) - 4\psi(0,0))/h^2 ,$$

and the increment h employed is

$$h = u^{1/6}\,\mathrm{ref}(\|x_*\|/\|v_1\|) .$$

The factor $u^{1/6}$ arises because the truncation error in the 9-point Laplacian is $O(h^4)$, and the roundoff error is $O(u/h^2)$: see Appendix D. Subroutine DIF3 calls DGFUN to evaluate expressions

$$2u_1^T \frac{\partial f}{\partial x}(x_* + (v_1 z + \bar{v}_1 \bar{z})/2 + w_2(z,\bar{z}))(v_1/2 + w_{20}z + w_{11}\bar{z}) ,$$

as functions of the variables $y_1 = \mathrm{Re}(z)$ and $y_2 = \mathrm{Im}(z)$. The user-supplied subroutine FNAME is called by DGFUN to evaluate the Jacobian matrix.

3.7. Computation of $c_1(0)$

The coefficient $c_1(0)$ of the cubic term in the Poincaré normal form is given by

$$c_1(0) = \frac{i}{2\omega_0}(g_{20}g_{11} - 2|g_{11}|^2 - |g_{02}|^2/3) + G_{21}/2 .$$

The computation of $c_1(0)$, once the critical value ν_c and the corresponding stationary point $x(\nu_c)$ are known, is organized by subroutine C1PNF. Once all of g_{20}, g_{11}, g_{02} and G_{21} have been obtained as described in Sections 3.1 through 3.6 above, C1PNF evaluates $c_1(0)$ using the given formula.

4. EVALUATION OF $\alpha'(\nu_c)$, $\omega'(\nu_c)$, μ_2, τ_2 AND β_2

Once $c_1(0)$ has been found, only $\alpha'(\nu_c)$ and $\omega'(\nu_c)$ are needed to evaluate μ_2, τ_2 and β_2. The derivative

$$\lambda_1'(\nu_c) = \alpha'(\nu_c) + i\omega'(\nu_c)$$

s approximated by the symmetric difference quotient,

$$\lambda_1'(\nu_c) \approx (\lambda_1(\nu_c + \Delta\nu) - \lambda_1(\nu_c - \Delta\nu))/2\Delta\nu ,$$

here

$$\Delta\nu = u^{1/3} \nu_{ref} ;$$

nd ν_{ref} is as in Section 2. The factor $u^{1/3}$ arises ecause the truncation error in the difference quotient is $)((\Delta\nu)^2)$ and the roundoff error is $O(u/\Delta\nu)$: see Appendix D.

The eigenvalue $\lambda_1(\nu_c + \Delta\nu)$ is evaluated by solving for $_*(\nu_c + \Delta\nu)$ by means of Newton's method, then using inverse teration for the eigenvalue. The eigenvalue $\lambda_1(\nu_c - \Delta\nu)$ is valuated similarly. Subroutine DEVAL1 computes the difference uotient, and calls EVAL1 to compute $\lambda_1(\nu_c \pm \Delta\nu)$.

Subroutine BIFOR2 organizes the computation of μ_2, τ_2 and $_2$. BIFOR2 calls ANUCRT to determine ν_c, $x(\nu_c)$, and $\omega(\nu_c)$, alls C1PNF to evaluate $c_1(0)$, calls DEVAL1 to evaluate $_1'(\nu_c)$, and then calculates

$$\mu_2 = -\mathrm{Re}(c_1(0))/\alpha'(\nu_c) ,$$

$$\tau_2 = -(\mathrm{Im}(c_1(0)) + \mu_2\omega'(\nu_c))/\omega(\nu_c) .$$

. ERROR ESTIMATION

There are errors in the computed values of μ_2, τ_2 and β_2 ue to the use of finite difference approximations to the erivatives f_{20}, f_{11}, G_{21}, and $\lambda_1'(\nu_c)$. (The use of a ifference approximation to start the iteration to locate ν_c , oes not in general affect the computed value of ν_c and may be gnored.) The schemes employed in approximating for f_{20}, f_{11}, $_{21}$ and $\lambda_1'(\nu_c)$ are all such that the total error due to ifferencing (truncation and roundoff) is $O(u^{2/3})$, u being the elative machine precision. Thus the error due to the use of differencing in the computed values of μ_2, τ_2 and β_2 is also $O(u^{2/3})$.

In order to provide information about the actual error involved, BIFOR2 will (when JJOB = 1 is specified) estimate the error due to the use of differencing. The computations of $c_1(0)$ and of $\lambda_1'(\nu_c)$ are performed both with increments based upon the user-supplied value u, and with increments based upon u increased by a factor of 1000 from the value input. Let err_4, err_5, err_6 and err_7 denote (respectively) the absolute values of the changes in the computed values of $\text{Re } c_1(0)$, $\text{Im } c_1(0)$, $\alpha'(\nu_c)$ and $\omega'(\nu_c)$. Since $\beta_2 = 2 \text{ Re } c_1(0)$, $err_3 = 2\, err_4$ is an estimate of the error due to differencing in β_2. The quantities μ_2 and τ_2 are then recomputed on the basis of values $\text{Re } c_1(0) \pm err_4$, $\text{Im } c_1(0) \pm err_5$, $\alpha'(\nu_c) \pm err_6$ and $\omega'(\nu_c) \pm err_7$, and the absolute values of the changes in the computed values of μ_2 and τ_2 are used as estimates err_1 and err_2 of the error due to differencing in μ_2 and τ_2.

6. SAMPLE APPLICATIONS

In this section, we give six applications of the code for evaluating bifurcation formulae. The last two applications are to partial differential systems.

The first example, the mass-spring-belt problem, was discussed both in Chapters 1 and 2. Here, the example serves as a simple introduction to use of the code.

The second example, Watt's centrifugal governor, is a third order system representing one of the oldest control systems known ([93], pp. 213-220). We present a criterion (in the form of a function $\rho = \rho_0(\kappa)$) for deciding whether the bifurcation that occurs when Vyshnegradskii's stability condition is violated is to stable periodic solutions or to unstable periodic solutions. Our study has design implications. In a control system in which there is a possibility of violating the stability condition (an elevator starting from rest, say) it would be good ractice to ensure that should the equilibrium state lose tability, the loss of stability is to stable, small amplitude eriodic solutions.

The third example is Lorenz' model for dynamic turbulence. e find that the bifurcation is to unstable periodic solutions or all values of the parameters considered. This agrees with arsden and McCracken's corrected version of their original nalysis [81; pp. 141-148], and with [104].

The fourth example is drawn from the study [44] of periodic olutions of the Hodgkin-Huxley model nerve equations. The resent technique, however, is simpler in that numerical ifferencing rather than symbolic manipulation is used to evalu-te the second and third partial derivatives. As in [44], we ind unstable periodic solutions for $I \approx I_1$, $I < I_1$, and stable eriodic solutions for $I \approx I_2$, $I < I_2$. A discussion of the lobal bifurcation diagram is presented.

The fifth example is the Lefever-Prigogine system (the russelator) with diffusion, introduced in Chapter 2 but here onsidered with fixed boundary conditions on the unit interval. he analytic theory of this partial differential system is resented in Chapter 5. In the present chapter, a simple finite ifference scheme is used to reduce the p.d.e. to an approxi-ating system of o.d.e.'s. The present results verify that the omplicated exact expressions in Chapter 5 are indeed correct.

The final example, the Dowell [30] plate flutter problem is aken from recent work of Holmes and Marsden [53]. The problem s actually a partial integro-differential system, which under alërkin expansion is reduced to an approximating system of rdinary differential equations. In [53], a 4 by 4 system was mployed. Here, we observe convergence of the various parameters ith increasing order of the approximating system up to 20 by 20. This confirms the work of Holmes and Marsden [53].) A nice icture of the mechanism of loss of stability is obtained.

Example 1. The mass-spring-belt system.

For the general discussion of this system, see Chapter 1; and for the exact evaluation of the bifurcation parameters μ_2, τ_2, β_2, see Chapter 2. Since the exact expressions are readily obtained, one would not normally compute them numerically except perhaps to verify the analytical expressions. The present computation is to be viewed as an introductory exercise in application of the code BIFOR2.

To use the code, it is necessary to specify a function $F(\nu)$. We chose

$$F(\nu) = \begin{cases} 0 & \text{if} \quad \nu = 0 , \\ 1 & \text{if} \quad \nu \geq 1 , \\ 1 + (1 - \nu)^3/3 & \text{if} \quad 0 < \nu < 1 , \\ -F(-\nu) & \text{if} \quad \nu < 0 . \end{cases} \qquad (6.1)$$

Subprograms SB, SBFUN, and SBCF were written to evaluate μ_2, τ_2 and β_2 for the values $m = 2$, $c = .1$, and $k = .5$. The calculation was repeated for different values of u to explore the effect of numerical differencing. The following table of values was obtained.

Table 3.1. Effect of numerical differencing.

u	μ_2	τ_2	β_2
10^{-8}	.09882117688	.004166594142	.03125000000
10^{-10}	.09882117688	.004166663301	.03125000000
10^{-12}	.09882117688	.004166666512	.03125000000
10^{-14}	.09882117688	.004166666676	.03125000000
10^{-16}	.09882117695	.004166666692	.03125000000
10^{-18}	.09882117712	.004166666787	.03125000001
10^{-20}	.09882117602	.004166667098	.03125000005

The exact values are $\mu_2 = 10^{\frac{1}{2}}/32 = .09882117688026...$,

$\tau_2 = .00416$ and $\beta_2 = .03125$.

The unexpected accuracy in the computed values of μ_2 and β_2 for large u occurs because $\mathfrak{F}'(\nu)$ is locally a quadratic. Consequently, the numerical differencing can compute α', $\mathfrak{F}''$ and $\mathfrak{F}'''$ exactly in the absence of roundoff error, and for large u roundoff error is negligible. The error in τ_2 for large u is due to truncation error in evaluating $\omega'(0)$. Since roundoff error is machine dependent, Table 3.1 for the smaller values of u reflects the fact that the computation was performed in single precision on a CDC Cyber 174, a machine which carries roughly 14 decimal digits (48 bit mantissa).

Example 2. The centrifugal governor.

A simplified version of Vyshnegradskii's analysis of a steam engine with Watt's centrifugal governor is presented by Pontryagin [93; pp. 213-220] in his textbook on ordinary differential equations.

Pontryagin's model system is

$$\frac{d\varphi}{d\tau} = \psi$$

$$\frac{d\psi}{d\tau} = n^2\omega^2 \sin\varphi\cos\varphi - \frac{g}{\ell}\sin\varphi - \frac{b}{m\ell}\psi \tag{6.2}$$

$$\frac{d\omega}{d\tau} = \frac{k}{J}\cos\varphi - \frac{F}{J},$$

where φ is the angle of deviation of the arms of the centrifugal governor from vertical, τ is the independent variable (time), ψ is defined by the first equation, ω is the angular velocity of the flywheel, $n\omega$ is the angular velocity of the governor, n is the constant transmission ratio, g is the acceleration due to gravity, b is a measure of the frictional force, ℓ is the length of each arm of the governor, and m is the mass of each weight suspended from the arms. The third equation of (6.2) arises from

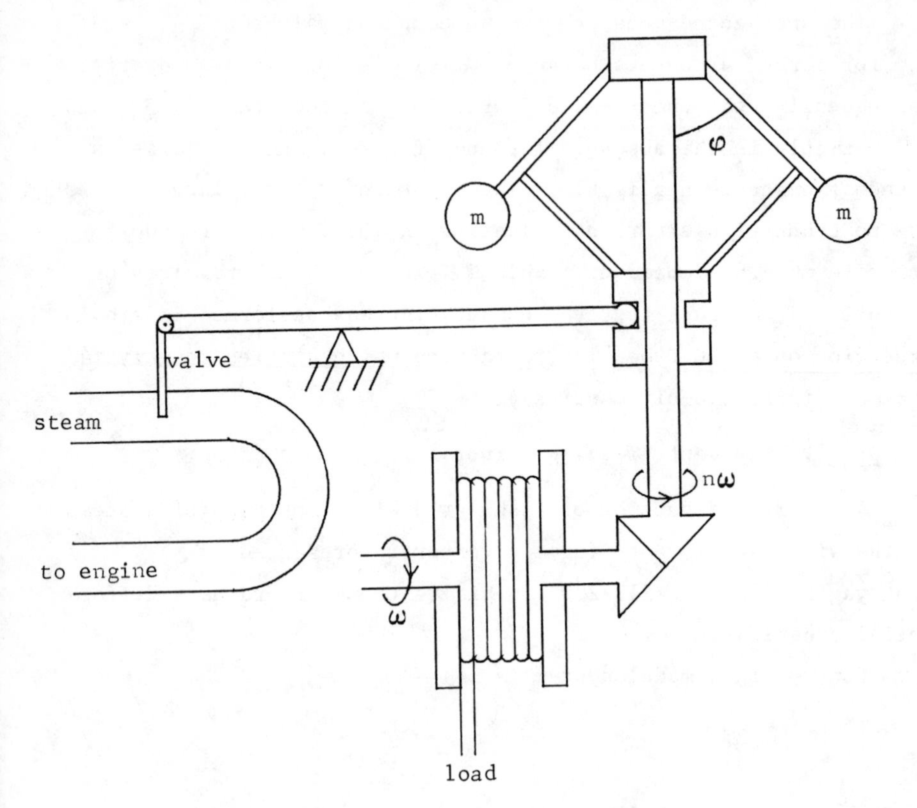

Figure 3.1. Steam engine centrifugal governor

$$J \frac{d\omega}{d\tau} = P_1 - P ,$$

where J is the moment of inertia of the flywheel, $P_1(\varphi)$ is the moment due to the action of the steam and P is the moment due to the load on the steam engine. As the angle φ varies, the vertical motion of the sleeve of the governor is a linear function of $\cos\varphi$, as is the motion of the valve which admits the steam. It is thus reasonable to write

$$P_1(\varphi) = P_1(\tilde{\varphi}) + k(\cos\varphi - \cos\tilde{\varphi}) ,$$

where $\tilde{\varphi}$ is an angle near which φ is to be maintained, and $k > 0$ is a proportionality constant. The constant F in equation (6.2) is then

$$F = P - P_1(\tilde{\varphi}) + k\cos\tilde{\varphi} ,$$

and the angle φ^* for which $d\omega/d\tau = 0$ is

$$\varphi^* = \cos^{-1}(F/k) .$$

To reduce the number of independent parameters we perform the change of variables

$$\varphi = x_1 , \quad \psi = \left(\frac{g}{\ell}\right)^{\frac{1}{2}} x_2 , \quad \omega = \frac{g^{\frac{1}{2}}}{n\ell^{\frac{1}{2}}} x_3 , \quad \tau = \left(\frac{\ell}{g}\right)^{\frac{1}{2}} t \qquad (6.3)$$

under which (6.2) becomes

$$\dot{x}_1 = x_2$$

$$\dot{x}_2 = (\sin x_1 \cos x_1) x_3^2 - \sin x_1 - \gamma x_2 \qquad (6.4)$$

$$\dot{x}_3 = \kappa(\cos x_1 - \rho) ,$$

where

$$\gamma = \frac{b}{m(\ell g)^{\frac{1}{2}}} , \quad \kappa = \frac{nk\ell}{Jg} , \quad \text{and} \quad \rho = \frac{F}{k} . \qquad (6.5)$$

In terms of the parameters γ , κ , ρ , Vyshnegradskii's stability condition is

$$\gamma > \gamma_c(\kappa,\rho) \equiv 2\kappa\rho^{3/2} . \qquad (6.6)$$

We shall now examine what happens when γ is decreased past γ_c so that the stationary point

$$x_* = (\cos^{-1}\rho , 0, \rho^{-\frac{1}{2}}) \qquad (6.7)$$

loses stability. The characteristic equation of the Jacobian matrix of (6.4) at the stationary point (6.7) is

$$\lambda^3 + \gamma\lambda^2 + \left(\frac{1}{\rho} - \rho\right)\lambda + 2\kappa\rho^{\frac{1}{2}}(1 - \rho^2) = 0$$

while the general cubic with real root λ_3 and pure imaginary complex conjugate roots $\lambda_{1,2} = \pm i\omega_0$ is

$$\lambda^3 - \lambda_3\lambda^2 + \omega_0^2\lambda - \lambda_3\omega_0^2 .$$

Comparing coefficients at $\gamma = \gamma_c = 2\kappa\rho^{3/2}$, we see that the roots of (6.8) are

$$\lambda_{1,2}(\gamma_c) = \pm i\omega_0 = \pm i\left(\frac{1}{\rho} - \rho\right)^{\frac{1}{2}} \quad \text{and}$$

$$\lambda_3(\gamma_c) = -\gamma_c .$$

Differentiation of (6.8) gives

$$\lambda'(\gamma) = -1/[3 + 2\gamma/\lambda + (\omega_0/\lambda)^2]$$

$$\lambda_1'(\gamma_c) = \bar{\lambda}_2'(\gamma_c) = -\frac{1}{2}\left(\frac{1 + i\gamma_c/\omega_0}{1 + (\gamma_c/\omega_0)^2}\right) .$$

The loss of stability at $\gamma = \gamma_c$ is thus a Hopf bifurcation, in which

$$\alpha'(0) = \operatorname{Re} \lambda_1'(\gamma_c) = -\frac{1}{2}\left(1 + (\gamma_c/\omega_0)^2\right)^{-1} < 0 .$$

The eigenvector v_1 corresponding to $\lambda_1(\gamma_c) = i\omega_0$ is then

$$v_1(\gamma_c) = (1, i\omega_0, i\kappa\rho^{\frac{1}{2}})^T .$$

Our object is the computation of $\mu_2 = \mu_2(\kappa,\rho)$ for $0 < \rho < 1$ and $\kappa > 0$.

Before proceeding to the numerics, we note that it is possible, in principle, to write out the expression for μ_2

explicitly. The matrix P is

$$P = \begin{bmatrix} 1 & 0 & 1 \\ 0 & -\omega_0 & -2\kappa\rho^{3/2} \\ 0 & -\kappa\rho^{\frac{1}{2}} & \omega_0/2\rho \end{bmatrix}$$

and the function $F(y) = P^{-1}f(x_* + Py)$ is given by

$$F^3 = \frac{2\kappa\rho^{\frac{1}{2}}}{\Delta}\begin{pmatrix} \rho\sin[2(\varphi^* + y_1 + y_3)]x_3^2(y_2,y_3) \\ -\sin(2\varphi^* + y_1 + y_3) + \omega_0\rho^{3/2} \end{pmatrix},$$

$$F^1 = -(\omega_0 y_2 + \gamma_c y_3) - F^3 \text{ , and}$$

$$F^2 = -\frac{1}{\Delta}\begin{pmatrix} \omega_0\sin[2(\varphi^* + y_1 + y_3)]x_3^2(y_2,y_3)/2 \\ -\omega_0\sin(\varphi^* + y_1 + y_3) + \omega_0\gamma_c(\omega_0 y_2 + \gamma_c y_3) \\ +4\kappa^2\rho^{5/2}\cos(\varphi^* + y_1 + y_3) - 4\kappa^2\rho^{7/2} \end{pmatrix},$$

where

$$\Delta = \omega_0^2 + 4\kappa^2\rho^3 \quad \text{and}$$

$$x_3(y_2,y_3) = \rho^{-\frac{1}{2}} - \kappa\rho^{\frac{1}{2}}y_2 + \omega_0 y_3/2\rho .$$

To write out the expression for μ_2, it is only necessary to follow Steps 6-8 of the Recipe-Summary in Chapter 2. The result will be a rather lengthy rational function of $\rho^{\frac{1}{2}}$, κ and ω_0. This computation would make a nice exercise in symbolic manipulation.

It is much simpler to use BIFOR2 to evaluate $\mu_2(\kappa,\rho)$. Subroutines CG and CGFUN were written to perform this task. The following table of values of $\mu_2(\kappa,\rho)$ is taken from program output.

Table 3.2. μ_2 for the centrifugal governor.

ρ \ κ	.1	.3	.5	.7	.9	1.1
.01	-.0001	-.0002	-.0004	-.0005	-.0007	-.0008
.05	-.0008	-.0025	-.0042	-.0059	-.0075	-.0092
.1	-.0024	-.0071	-.0118	-.0165	-.0213	-.0260
.2	-.0065	-.0196	-.0327	-.0459	-.0593	-.0728
.3	-.0115	-.0346	-.0579	-.0816	-.1058	-.1307
.4	-.0166	-.0499	-.0837	-.1184	-.1542	-.1917
.5	-.0206	-.0622	-.1047	-.1487	-.1950	-.2444
.6	-.0218	-.0659	-.1117	-.1604	-.2137	-.2733
.7	-.0159	-.0492	-.0873	-.1336	-.1916	-.2646
.8	.0095	.0190	.0046	-.0394	-.1141	-.2204
.9	.1131	.2501	.2455	.1535	.0120	-.1690
.95	.3294	.5795	.4770	.2916	.0823	-.1542
.99	1.5779	1.2536	.7560	.4196	.1358	-.1521

The corresponding tables for $\tau_2(\kappa,\rho)$ and $\beta_2(\kappa,\rho)$ may be found on the microfiche.

From Table 3.2 we infer the existence of a function $\rho = \rho_0(\kappa)$ such that $\mu_2(\kappa,\rho_0(\kappa)) = 0$. Further computations produced the following table of the function $\rho_0(\kappa)$.

Table 3.3. $\rho_0(\kappa)$ such that $\mu_2(\kappa, \rho_0(\kappa)) = 0$.

κ	$\rho_0(\kappa)$	$\varphi_0(\kappa) = \cos^{-1} \rho_0(\kappa)$
.01	.7746	39.23
.05	.7748	39.21
.1	.7754	39.16
.2	.7778	38.94
.3	.7819	38.57
.4	.7881	37.99
.5	.7968	37.17
.6	.8089	36.01
.7	.8259	34.32
.8	.8507	31.71
.9	.8914	26.95
.95	.9260	22.18
.99	.9764	12.48
1.0	1.0000	0.00

For $\kappa > 1$, our computations indicate that $\mu_2(\kappa, \rho) < 0$ for all values of ρ , $0 < \rho < 1$.

Table 3.3 provides a simple criterion for deciding the direction of bifurcation . If $\kappa \geq 1$, or if $\kappa < 1$ and $\rho < \rho_0(\kappa)$, then $\mu_2(\rho, \kappa) < 0$ and for γ slightly smaller than $\gamma_c = 2\kappa\rho^{3/2}$ the system (6.4) has a stable, ($\beta_2 = -2\alpha'(0)\mu_2 < 0$) small amplitude oscillation

$$x(t) = x_* + \left(\frac{\gamma - \gamma_c}{\mu_2}\right)^{\frac{1}{2}} \mathrm{Re}(\exp(2\pi it/T)v_1) + O(\gamma - \gamma_c) .$$

Thus, although the stationary point becomes unstable, the system is still relatively well behaved. If $\kappa < 1$ and $\rho > \rho_0(\kappa)$, then $\mu_2(\rho, \kappa) > 0$; and there is an unstable family of oscillations for γ slightly larger than γ_c . For these values of γ , even though the stationary point is asymptotically stable, it is unstable with respect to certain perturbations of amplitude

$$\left(\frac{\gamma - \gamma_c}{\mu_2}\right)^{\frac{1}{2}} \|v_1\| + O(\gamma - \gamma_c) .$$

If $\rho = \rho_0(K)$, then $\mu_2(\rho,K) = 0$ and our results do not determine the direction of bifurcation. See Exercise 8 at the end of this Chapter.

An interesting feature of Table 3.3 is the lower bound for $\rho_0(K)$, or equivalently, the upper bound for $\varphi_0(K)$. We have

$$\varphi_0(K) \le \varphi_m \approx 39.3^\circ .$$

This bound implies the following: If the parameters ρ, K and γ are varied in any manner such that either $K > 1$ or $\varphi^* = \cos^{-1}\rho > \varphi_m \approx 39.3^\circ$, then any loss of stability of the stationary point that may occur is a Hopf bifurcation to stable periodic solutions.

Given a control system that may at times operate near a limit of stability of an equilibrium state, it is clearly good practice to adjust the control parameters so that any loss of stability that may occur is to stable rather than unstable periodic solutions. In the present example, we have shown how this may be accomplished for Pontryagin's model.

<u>Example 3</u>. The Lorenz system.

The system of equations

$$\begin{aligned} y_1' &= -\sigma y_1 + \sigma y_2 \\ y_2' &= -y_1 y_3 + \gamma y_1 - y_2 \qquad (6.9) \\ y_3' &= y_1 y_2 - b y_3 \end{aligned}$$

was studied by Lorenz (see [81], pp. 141-148) as a model for fluid dynamic turbulence. The system remains interesting, even though it has little to do with some current theories of turbulence.

In the Lorenz system, bifurcation with respect to the parameter γ is considered for fixed values of σ and b.

If $\sigma \neq 0$ and $b(\gamma - 1) > 0$, the stationary points of (6.9) are

$$y_1 = 0 , \quad y_1 = \pm[b(\gamma - 1)]^{1/2}$$
$$y_2 = y_1 \quad \text{and} \quad y_3 = y_1^2 = y_1^2/b . \tag{6.10}$$

The point corresponding to the positive square root is of interest.

The Jacobian of the system at this stationary point is

$$\begin{pmatrix} -\sigma & \sigma & 0 \\ 1 & -1 & -y_1 \\ y_1 & y_1 & -b \end{pmatrix} , \tag{6.11}$$

where $y_1 = [b(\gamma - 1)]^{1/2}$, and has characteristic polynomial

$$\lambda^3 + (\sigma + b + 1)\lambda^2 + b(\gamma + \sigma)\lambda + 2b\sigma(\gamma - 1) . \tag{6.12}$$

The general real cubic with real root λ_3 and pure imaginary roots $\lambda_{1,2} = \pm i\omega_0$ is

$$\lambda^3 - \lambda_3\lambda^2 + \omega_0^2\lambda - \lambda_3\omega_0^2 .$$

At the critical value of γ, we must therefore have

$$\lambda_3 = -(\sigma + b + 1) ,$$

$$\lambda_{1,2} = \pm i\omega_0 = \pm i\,[b(\gamma_c + \sigma)]^{1/2}$$

and

$$2b\sigma(\gamma_c - 1) = b(\gamma_c + \sigma)(\sigma + b + 1) , \text{ or}$$

$$\gamma_c = \frac{\sigma(\sigma + b + 3)}{\sigma - b - 1} . \tag{6.13}$$

Thus

$$b(\gamma_c + \sigma) = 2b\sigma(\sigma + 1)/(\sigma - b - 1) ,$$

which is positive if $\sigma > b + 1$. (We only consider $b > 0$, $\sigma > 0$.)

Differentiating (6.12) with respect to γ gives

$$\lambda'(\gamma) = -b(\lambda + 2\sigma)/ [3\lambda^2 + 2(\sigma + b + 1)\lambda + b(\gamma + \sigma)] ,$$

so

$$\alpha'(0) = \mathrm{Re}\ \lambda_1'(\gamma_c) = \frac{b(\sigma - b - 1)}{2[\omega_0^2 + (\sigma + b + 1)^2]} > 0 .$$

The loss of stability at $\gamma = \gamma_c$ given by (6.13) is thus a classical Hopf bifurcation. The eigenvector v_1 corresponding to $\lambda_1(\gamma_c) = i\omega_0)$ is

$$v_1(\gamma_c) = \left(1,\ 1 + i\frac{\omega_0}{\sigma},\ \frac{\omega_0}{y_1(\gamma_c)}\left[\frac{\omega_0}{\sigma} - i\left(1 + \frac{1}{\sigma}\right)\right]\right)^T .$$

Again, it is possible, in principle, to write out the expression for μ_2 explicitly. The task would make another nice exercise in symbolic manipulation.

Subroutines LR and LRFUN were written to evaluate μ_2, τ_2 and β_2 numerically. The following table of values of $\mu_2(\sigma,b)$ is taken from program output. The corresponding tables for $\tau_2(\sigma,b)$ and $\beta_2(\sigma,b)$ may be found on the microfiche.

Table 3.4. $\mu_2(\sigma,b)$ for the Lorenz system.

b \ σ	20	40	60	80	100
10.0	-.0399	-.0262	-.0238	-.0219	-.0202
20.0		-.0175	-.0134	-.0125	-.0121
30.0		-.0285	-.0111	-.0091	-.0085
40.0			-.0136	-.0082	-.0069
50.0			-.0276	-.0091	-.0064
60.0				-.0130	-.0069
70.0				-.0274	-.0086

For all values of σ and b that we consider, we find $\mu_2(\sigma,b) < 0$, i.e. the bifurcation is to unstable ($\beta_2 = -2\alpha'(0)\mu_2 > 0$) periodic solutions for $\gamma \approx \gamma_c$, $\gamma < \gamma_c$.

Remark 1. An analytical study of Hopf bifurcation in the Lorenz system is given in [81; pp. 141-148], and it was Jerry Marsden who suggested we consider the system. Our study revealed a mistake in Figure 4B.1 of [81]. See [104] for more recent analytical work on Hopf bifurcation in the Lorenz system.

Remark 2. In the Lorenz system, perhaps more interesting than the Hopf bifurcation is the presence of a strange attractor; see [81, pp. 368-381].

Example 4. The Hodgkin-Huxley current clamped system.

In the early 1950's a series of experiments was performed on the giant axon of the squid "Loligo" by Hodgkin, Huxley, and coworkers. One result was the famous model of nerve conduction which continues to play a central role in the theory of the nerve [52, 96]. The Hodgkin-Huxley model, in the general case, is represented by a partial differential system which governs the propogation of nerve impulses. There are two independent variables, time, and distance along the axon.

Solutions of the partial differential system which are functions of a single traveling wave variable may be obtained as solutions of a 5th order ordinary differential system. One of the successes of the Hodgkin-Huxley model is the prediction of the experimentally observed wavespeed from numerical solutions of this 5th order system.

Solutions of the partial differential system which are functions of time alone, independent of space, may be obtained as solutions of a 4th order ordinary differential system. Such solutions correspond to "space clamped" experiments in which a thin platinized-platinum wire (electrode) is inserted along the length of the axon. Because the platinized-platinum wire is a good conductor, the electric potential (voltage) at every point along the wire is essentially the same at any one time; this is assumed to eliminate all spatial dependence. In voltage clamped experiments, an external constant voltage source is connected to the electrodes (the other electrode being in the bath), and the resulting current is measured as a function of time. In current clamped experiments, an external constant current source is connected to the electrodes and the resulting voltage is measured as a function of time.

In the current clamped experiments, an interesting threshold phenomenon is observed. For small values of the external current stimulus, the voltage simply decays to a rest value. For slightly larger values of the current I, the nerve axon responds repetitively, with a number of large amplitude "action potentials", before the voltage decays to rest. Numerical integrations of the Hodgkin-Huxley model current clamped system reveal much the same phenomenon, except that the succession of action potentials continues forever as a periodic solution. The fact that the Hodgkin-Huxley model does not predict that the axon will eventually become "tired out" in the current clamped experiments is an inadequacy of the model. It does seem reasonable, however, to suppose that periodic solutions of the model correspond to repetitive firing of the axon. We note that in a modified experimental setting with a low calcium solution, the axon can be made to produce long sequences of action potentials.

In what follows, we shall use bifurcation theory to describe two different families of small amplitude periodic solutions of the model current clamped system. The system is given by

$$C_M \frac{dv}{dt} = G(v,m,n,h) + I$$

$$\frac{dm}{dt} = [(1 - m)\alpha_m(v) - m\beta_m(v)]\Phi$$

$$\frac{dn}{dt} = [(1 - n)\alpha_n(v) - n\beta_n(v)]\Phi$$

$$\frac{dh}{dt} = [(1 - h)\alpha_h(v) - h\beta_h(v)]\Phi ,$$

where v is the voltage in mv, the time t is measured in msec, m is the sodium activation, n is the potassium activation, h is the sodium inactivation, and $\Phi = 3^{(T-6.3)/10}$ is a temperature compensation factor. The temperature T is measured on ^oC, and the factor Φ alters the reaction rate "constants" $\alpha_m(v)$, $\beta_m(v)$, $\alpha_n(v)$, $\beta_n(v)$, $\alpha_h(v)$ and $\beta_h(v)$. Explicitly, Hodgkin and Huxley give

$$\alpha_m(v) = 1/\mathrm{expc}(-.1v + 2.5), \quad \beta_m(v) = 4\exp(-v/18)$$

$$\alpha_n(v) = .1/\mathrm{expc}(-.1v + 1), \quad \beta_n(v) = \exp(-v/80)/8 ,$$

$$\alpha_h(v) = .07\mathrm{expc}(-.05v), \quad \beta_h(v) = 1/(1 + \exp(-.1v + 3)) ,$$

where

$$\mathrm{expc}(x) = \begin{cases} (e^x - 1)/x & \text{if } x \neq 0 , \\ 1 & \text{if } x = 0 . \end{cases}$$

The function $G(v,m,n,h)$ is $G =$
$120\, m^3 h(115 - v) - 36n^4(v + 12) + .3(c_{10} - v)$, where the constant c_{10} is adjusted so that $G(0, m_\infty(0), n_\infty(0), h_\infty(0)) = 0$. The subscript ∞ indicates rest values:

$$m_\infty(v) = \alpha_m(v) / (\alpha_m(v) + \beta_m(v)) ,$$

$$n_\infty(v) = \alpha_n(v) / (\alpha_n(v) + \beta_n(v)) ,$$

$$h_\infty(v) = \alpha_h(v) / (\alpha_h(v) + \beta_h(v)) .$$

One finds $c_{10} \approx 10.599$. C_M is the membrane capacitance per unit area, $C_M = 1.0\ \mu f/cm^2$, and I is the external current stimulus. Both I and G are measured in μA .

For each value of the parameter I , there is a unique stationary point $x_*(I) = (v_*(I), m_*(I), n_*(I), h_*(I))^T$. For reasonable temperatures T , there are two critical values of I , $I_1 \approx 10$ and $I_2 \approx 150$ at which the system exhibits a Hopf bifurcation. The presence of these Hopf bifurcations is revealed by linear stability analysis of the stationary point, see [22, 44].

Subprograms HH, HHFUN and HHEXP were written to evaluate μ_2, τ_2 and β_2 at I_1 and I_2 for temperatures $T = 0^\circ C$ and $T = 6.3^\circ C$. The results are given in the following table.

<u>Table 3.5</u>. μ_2, τ_2 and β_2 for the Hodgkin-Huxley system.

$T = 0^\circ C$	$I_1 = 8.42$	$\mu_2 = -.0833$	$\tau_2 = .0149$	$\beta_2 = .0024$
$T = 0^\circ C$	$I_2 = 152.30$	$\mu_2 = -.2711$	$\tau_2 = .00050$	$\beta_2 = -.0014$
$T = 6.3^\circ C$	$I_1 = 9.78$	$\mu_2 = -.1154$	$\tau_2 = .0114$	$\beta_2 = .0043$
$T = 6.3^\circ C$	$I_2 = 154.53$	$\mu_2 = -.2799$	$\tau_2 = .00045$	$\beta_2 = -.0025$

In all cases, $\mu_2 < 0$ so the periodic solutions exist for $I < I_j$ $(j = 1,2)$. The periodic solutions for $I < I_1$ are unstable $(\beta_2 > 0)$ while those for $I < I_2$ are stable.

The families of periodic solutions found for $I < I_1$, $I \approx I_1$ and for $I < I_2$, $I \approx I_2$ by means of Hopf bifurcation theory are actually the two "ends" of a single, larger family of periodic solutions as shown in Figure 3.2.

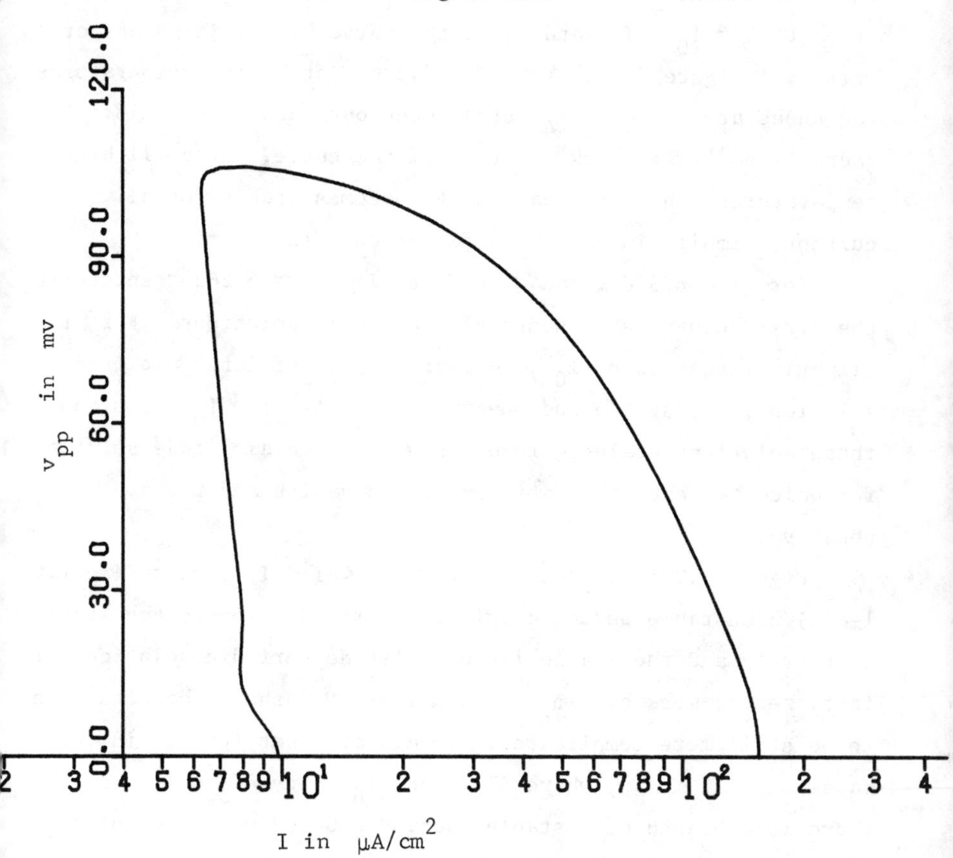

Figure 3.2. Bifurcation diagram for the current-clamped Hodgkin-Huxley system, $T = 6.3^\circ C$.

Each point (I, v_{pp}) on the curve represents a periodic solution of the current-clamped Hodgkin-Huxley system for a current stimulus I, and of amplitude measured by the peak-to-peak voltage

$$v_{pp} \equiv \max_t v(t) - \min_t v(t) .$$

Figure 3.2 was constructed by purely numerical techniques,

and includes work of Rinzel and Miller [96]. For $I \approx I_1$ and $I \approx I_2$, the numerical results agree with the bifurcation analytic predictions in precisely the way expected; see [44] for the comparison.

At $I = I_0$, I_3 and I_4, the curve has limit points or "knees". Figure 3.2 is for $T = 6.3^{\circ}C$: at higher temperatures, the knees at I_3 and I_4 straighten out, and at $T = 18.5^{\circ}C$, there is no "switchback" portion of the curve. At still higher temperatures, the stationary state becomes stable for all current stimuli I, and the bifurcations disappear.

For $T = 6.3^{\circ}C$, the "knee" at $I = I_0 \approx 6.265$ represents the first occurence of nontrivial periodic solutions. For I slightly larger than I_0, there is a pair of large amplitude solutions, one stable and one unstable. As $I \to I_0$ from above, these solutions coalesce into a single large amplitude solution for which two characteristic exponents vanish and two are negative.

Figure 3.2 indicates that for $I_3 < I < I_4$, there are (at least) 3 unstable periodic solutions, in addition to the stable rest state and the stable large amplitude periodic solution. In fact, recent work by Rinzel and Miller [96] shows the situation to be still more complicated. It appears that for I in the range $I_5 < I < I_6$, where I_5 and I_6 obey $I_3 < I_5 < I_6 < I_4$, there is a branch of unstable, period-doubled periodic solutions. The period-doubled solutions have voltage peaks which are alternately larger and smaller. These solutions arise by means of secondary bifurcations at I_5 and I_6 in which a characteristic multiplier associated with the basic periodic solution passes through the point -1. At present writing, the full branch of period-doubled solutions has not been computed and the possibility of tertiary bifurcations has not been excluded. It seems to be a fairly safe conjecture, however, that no additional stable periodic solutions will be found.

Of the periodic solutions of the current-clamped

Hodgkin-Huxley system discussed for $I_0 < I < I_1$, only the rest state and the largest amplitude periodic solution can be expected as long-time behavior of the system because the periodic solutions of intermediate amplitudes are unstable, as indicated by computations of the associated characteristic multipliers.

Physically, the most relevant unstable periodic solutions are the small amplitude solutions for $I \approx I_1$, $I < I_1$. These solutions are analogous to the unstable limit cycles in the mass-spring-belt system (Example 1) in that they represent a mechanism by which perturbations from the rest state either decay back to rest or grow as $t \to \infty$, presumably to the largest amplitude solution.

For I fixed, $I \approx I_1$, $I < I_1$, the unstable small amplitude periodic solution is an unstable limit cycle in the two dimensional slice I = constant of the center manifold. If trajectories are begun with initial conditions on this two-dimensional invariant manifold and in the interior of the set bounded by the limit cycle, the trajectories will decay in an oscillatory manner to the rest state. If the initial conditions are on the invariant manifold but in the exterior of this same set, the trajectories oscillate with increasing amplitude. The unstable periodic solution also lies on a 3-dimensional hypersurface in $\mathbb{R}^4$, which defines a boundary for the stable manifold associated with the stable rest state $x_*(I)$, $I \approx I_1$, $I < I_1$. The set

$$\{x_0 \mid |2u_1^T(x_0 - x_*(I))|^2 < (I - I_1)/\mu_2, \ |x_0 - x_*(I)| < \delta\}$$

where δ is small but independent of I, is a local approximation to the stable manifold. For u_1, see Section 3.2.

Remarks.

1. It was the Hodgkin-Huxley current-clamped system which inspired the derivation of bifurcation formulae [46] upon which Chapters 1 and 2 are based. The study [44] initiated the

development of the computer program described in the present Chapter.

2. In Hassard [44], the evaluation of 2nd and 3rd order partial derivatives was by means of symbolic manipulation and equations (4.3) and (4.4). The present computation using the program for evaluation of bifurcation formulae is simpler, in that explicit expressions for the 2nd and 3rd order partials are not required.

3. Simple shooting was employed to compute most of Figure (3.2), with the variable order, variable step size Gear stiff system solver used for the individual shots; see [44].

As described in [44], however, the simple shooting scheme was unable to compute the full branch of unstable periodic solutions because of the extreme sensitivity of the trajectories with respect to initial conditions. The primary branch of unstable periodic solutions was completed by Rinzel and Miller [96] with a finite difference scheme using Gear's 5th order stiffly stable formula to discretize the time derivatives on a uniform mesh. The resulting large systems of nonlinear algebraic equations were solved by Newton's method, with special techniques to take advantage of the band structure of the Jacobian matrices.

Example 5. The Brusselator with fixed boundary conditions

Consider the system

$$\frac{\partial x}{\partial t} = d\,\frac{\partial^2 x}{\partial r^2} + (B - 1)x + A^2 y + h(x,y)$$

$$\frac{\partial y}{\partial t} = \theta d\,\frac{\partial^2 y}{\partial r^2} - Bx - A^2 y - h(x,y) \ ,$$

where

$$h(x,y) = BA^{-1}x^2 + (x + 2A)xy \ ,$$

and $x = y = 0$ at $r = 0,1$. The present version is obtained by setting $X = x + A$, $Y = y + B/A$, $D_1 = d$ and $D_2 = \theta d$ in the version discussed in Chapter 2 (Example 3, p. 101).

To convert this system to an approximating, ordinary differential system we use a simple finite difference scheme. After choosing an integer $n \geq 1$ we let $N = 2n$, $\Delta r = 1/(n + 1)$ and $r_i = i\Delta r$, for $i = 0,\ldots,n + 1$. Approximating $x(r_i)$ and $y(r_i)$ by x_i and y_i respectively $(i = 0,\ldots,n + 1)$ and using 3-point centered difference approximations for $\partial^2x/\partial r^2$, $\partial^2y/\partial r^2$, we obtain the N-dimensional ordinary differential system

$$\dot{x}_i = D(x_{i+1} + x_{i-1} - 2x_i) + (B - 1)x_i + A^2y_i + h(x_i,y_i)$$

$$\dot{y}_i = \theta D(y_{i+1} + y_{i-1} - 2y_i) - Bx_i - A^2y_i - h(x_i,y_i)$$

$$(i = 1,\ldots,n),$$

where $D = d/(\Delta r)^2$ and $x_0 = y_0 = x_{n+1} = y_{n+1} \equiv 0$ for all t. Subprograms BD and BDFUN were written to evaluate bifurcation formulae for the approximating system.

To investigate the effect of the finite difference scheme, we set $A = 1$, $d = .1$, $\theta = .5$ and computed B_c, μ_2, τ_2 and β_2 for $N = 2, 6, 14, 30$. Before one compares these results, however, it is necessary to standardize the normalizations.

Suppose $u(r,t)$ and $v(r,t)$ are smooth functions of r and t for $0 \leq r \leq 1$ and $-\infty < t < \infty$, are T periodic in t, and satisfy $u(v,t) = u(1,t) = v(0,t) = v(1,t) = 0$. A convenient norm for the pair (u,v) is

$$\|(u,v)\|^2 \equiv \frac{1}{T}\int_0^T [\int_0^1 u(r,t)^2 + v(r,t)^2 dr]dt .$$

Now by the trapezoidal rule,

$$\|(u,v)\|^2 = \frac{1}{T}\int_0^T [\Delta r \sum_{i=1}^{n} [u(r_i,t)^2 + v(r_i,t)^2] dt + O((\Delta r)^2) .$$

In the present application such functions $(u,v) = (x(r,t), y(r,t))$ are approximated using a vector function

$$X(t) = (x_1, x_2, \dots, x_n, y_1, y_2, \dots, y_n)^T$$

which, from Hopf theory, expands as

$$X = \left(\frac{\nu - \nu_c}{\mu_2}\right)^{1/2} \mathrm{Re}(e^{2\pi i t/T} v_1) + O(\nu - \nu_c) .$$

Therefore

$$\frac{1}{T}\int_0^T \Delta r \sum_{i=1}^{n} [x_i(t)^2 + y_i(t)^2]\, dt = \left(\frac{\nu - \nu_c}{\mu_2}\right) \frac{\Delta r}{2} |v_1|^2 + O(|\nu - \nu_c|^{3/2}),$$

and it is evident that

$$\bar{\mu}_2 \equiv \frac{2\mu_2}{\Delta r |v_1|^2}$$

should be compared for increasing N rather than μ_2 .

Similarly, define $\bar{\tau}_2 = 2\tau_2/(\Delta r|v_1|^2)$, $\bar{\beta}_2 = 2\beta_2/(\Delta r|v_1|^2)$. These quantities $\bar{\mu}_2$, $\bar{\tau}_2$, $\bar{\beta}_2$ are the values for μ_2, τ_2 and β_2 which arise when the eigenvector v_1 is normalized such that $\frac{\Delta r}{2} |v_1|^2 = 1$.

With $A = 1$, $d = .1$ and $\theta = .5$, we computed B_c, $\bar{\mu}_2$, $\bar{\tau}_2$ and $\bar{\beta}_2$ for $N = 2, 6, 14$ and 30 . The results form Table 3.6.

Table 3.6. Effect of increasing N ($A = 1$, $d = .1$, $\theta = .5$).

N	Δr	B_c	$\bar{\mu}_2$	$\bar{\tau}_2$	$\bar{\beta}_2$
2	.50000	3.200000	1.046083	.033241	-1.046083
6	.25000	3.405887	.768911	.017429	-.768911
14	.12500	3.461513	.763703	.015498	-.763703
30	.06250	3.475690	.762614	.015254	-.762614

We extrapolated these results using Richardson extrapolation, based upon assumed functional dependence of the form $f(\Delta r) = f(0) + c_2(\Delta r)^2 + c_3(\Delta r)^3 + O((\Delta r)^4)$. This procedure resulted in

$$B_c \approx 3.4804 \ , \quad \bar{\mu}_2 \approx .7623 \ , \quad \bar{\tau}_2 \approx .0152 \ , \quad \bar{\beta}_2 \approx -.7623$$

which may be compared with the exact values $B_c = 3.48044..$, $\bar{\mu}_2 = .76226..$, $\bar{\tau}_2 = .01518..$, $\bar{\beta}_2 = -.76226..$ from Chapter 5.

The effect of differing values of A, d and θ is investigated in Chapter 5.

Remarks.

1. Both the centered difference approximation to the 2nd partial derivatives and the trapezoidal rule approximation to the normalization condition have truncation errors of the form const $(\Delta r)^2 + O(\Delta r^3)$. If a higher order difference scheme is adopted, a corresponding higher order quadrature scheme should also be adopted.

2. Since BIFOR2 is designed for o.d.e. applications, it takes little advantage of the sparseness of the matrices involved in the present bifurcation calculation. To take full advantage of this sparseness, one could write a code similar to BIFOR2 but intended for one space-dimensional partial differential systems: such a code would use a nonlinear two-point boundary value problem solver and an ordinary differential eigenvalue problem solver. See Chapter 5, Exercise 7 (p. 265).

3. A simple way to 'get the most' out of BIFOR2 in applications to p.d.e. is to use very high order discretization schemes, since BIFOR2 was designed to handle dense matrices.

Example 6. A panel flutter problem.

This problem, which also involves a partial differential system, is included here to illustrate another way the program for evaluation of bifurcation formulae for systems of o.d.e.'s may actually be applied to p.d.e.'s.

The original problem is that of 'one-dimensional' panel flutter [30, 53]. We thank P. Holmes for suggesting the problem and for much of the following discussion.

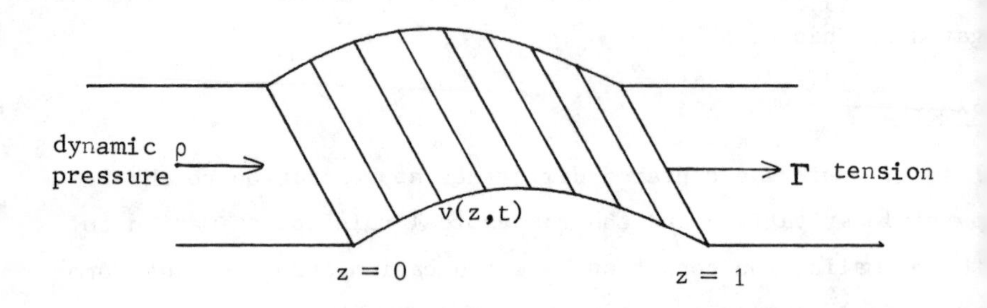

Figure 3.3. Panel in wind tunnel

If spanwise bending is negligible, then one can reduce the von Karman equations to an Euler type beam equation with flow effects derived from simple "piston theory":

$$\ddot{v} + \sqrt{\rho}\delta\dot{v} + \rho v' + \alpha\dot{v}'''' + v''''$$

$$= \{\Gamma + \varkappa\int_0^1 v'(z)^2 dz + \sigma\int_0^1 \dot{v}'(z)v'(z)dz\}v'' , \qquad (6.14)$$

where

$\rho v'$ = energy from flow,

$\sqrt{\rho}\,\delta\dot{v}$ = 'flow' damping,

Γ = axial tension,

$\varkappa$ = (membrane) stiffness,

α,σ = damping coefficients of panel,

with 'hinged' boundary conditions

$$v = v'' = 0 \quad \text{at} \quad z = 0,1 .$$

The finite dimensional Galërkin approximation is obtained, for m modes of vibration, by setting

$$v(z,t) = \sum_{j=1}^{m} x_j(t)\sin(j\pi z) \tag{6.15}$$

in (6.14) and carrying out the standard Galërkin method of taking the product with $\sin i\pi z$ and integrating from $z = 0$ to $z = 1$

This procedure produces the system

$$\frac{d}{dt}\begin{pmatrix} x_1 \\ x_m \\ \dot{x}_1 \\ \dot{x}_m \end{pmatrix} = \left[\begin{array}{c|c} 0 & I \\ \hline C & D \end{array}\right]\begin{pmatrix} x_1 \\ x_m \\ \dot{x}_1 \\ \dot{x}_m \end{pmatrix} + \begin{pmatrix} 0 \\ 0 \\ H_1 \\ H_m \end{pmatrix}, \tag{6.16}$$

where C, D and H are

$$C_{ij} = -\delta_{ij}(i\pi)^2\,[\Gamma + (i\pi)^2] - 2\rho j\pi \int_0^1 \sin i\pi z \cos j\pi z \, dz$$

$$D_{ij} = -\delta_{ij}[(i\pi)^4\alpha + (\rho)^{1/2}\delta] \quad (i,j = 1,\ldots,m) \quad \text{and} \tag{6.17}$$

$$H_i = [\sum_{j=1}^{m} (j\pi)^2(\kappa x_j + \sigma\dot{x}_j)x_j]\,\frac{(i\pi)^2 x_i}{2} \quad (i = 1,\ldots,m) .$$

With the definitions $N = 2m$ and $y \equiv (x_1,\ldots,x_m,\dot{x}_1,\ldots,\dot{x}_m)^t$, the original p.d.e. is then approximated by an N dimensional o.d.e. for the vector y .

The parameters κ , σ , α , δ and Γ are fixed, and bifurcation with respect to the parameter ρ is considered.

For the values

$$\varkappa = .01 \ , \quad \sigma = .0001$$

$$\alpha = .005 \ , \quad \delta = .1 \ , \quad \text{and} \quad \Gamma = -2.4\pi^2 \ , \tag{6.18}$$

Holmes has computed that, for $N = 8$, there is a critical value of $\rho \approx 112.75$, with stationary point

$$y_* \approx (2.4 \ , -1.04 \ , .094 \ , -.017 \ , 0 \ , 0 \ , 0 \ , 0) \ .$$

Since the number of modes of vibration is to be increased for a better approximation of the p.d.e., the code was written for variable (even) dimension N. Subroutines PF and PFFUN were written to use Holmes' estimates of ρ_c and of $y_*(\rho_c)$ for $N = 8$. For successively higher dimensions N, the converged values of ρ_c and $y_*(\rho_c)$ for the previous dimension N were used as estimates. The following set of values of μ_2, τ_2, β_2 were obtained as N was increased.

Table 3.7. μ_2, τ_2 and β_2 for the panel flutter problem.

N	ρ_c	μ_2	τ_2	β_2
8	112.778	-.058007	.089784	.019916
12	112.841	-.057878	.091871	.019806
16	112.846	-.057865	.092087	.019794
20	112.846	-.057862	.092130	.019791

In this example, the normalization provided by BIFOR2 for the eigenvector v_1 happens to be appropriate for the approximation scheme, so the values of μ_2, τ_2 and β_2 for increasing N may be compared immediately. Similar convergence of the stationary point and of the eigenvector v_1 was observed as N was increased,

$$y_* \to (2.2976, -.9995, .0906, -.0161, .0041, -.0018, .0007, \ldots$$

$$v_1 \to (1, -.4145 + .0084i, .0370 - .0012i, -.0069 + .0001i, .0017 - .0001i, \ldots .$$

Since $\mu_2 < 0$ and $\beta_2 > 0$, the bifurcation is subcritical and to unstable periodic solutions. Now the first m components of

$$y(t;\rho) = y_*(\rho_c) + \left(\frac{\rho - \rho_c}{\mu_2}\right)^{\frac{1}{2}} \operatorname{Re}(e^{2\pi it/T} v_1) + O(\rho - \rho_c)$$

provide approximations to the Fourier sine coefficients of $v(z,t;\rho)$. Thus

$$v(z,t;\rho) \approx v_*(z;\rho_c) + \left(\frac{\rho - \rho_c}{\mu_2}\right)^{\frac{1}{2}} \left[\cos \frac{2\pi t}{T} C(z) - \sin \frac{2\pi t}{T} S(z)\right] ,$$

where

$$v_*(z;\rho) = \sum_{j=1}^{m} (x_*(\rho))_j \sin j\pi z ,$$

$$C(z) = \sum_{j=1}^{m} (\operatorname{Re} v_1)_j \sin j\pi z ,$$

$$S(z) = \sum_{j=1}^{m} (\operatorname{Im} v_1)_j \sin j\pi z ,$$

and the approximation to $v(z,t;\rho)$ contains error due to truncation of the Fourier series in addition to the usual error $O(\rho - \rho_c)$. The functions $v_*(z;\rho_c)$, $C(z)$ and $S(z)$ obtained for $m = 10$ are shown in Figure 3.4. The function v_* is the base profile, buckled due to the combined effects of compression (negative tension) and dynamic pressure. The base profile loses linear stability as ρ increases past ρ_c. For $\rho \approx \rho_c$, $\rho < \rho_c$ even though the base solution is asymptotically stable, it is unstable with respect to certain perturbations of the form

$$\left(\frac{\rho - \rho_c}{\mu_2}\right)^{\frac{1}{2}} \left[\cos \frac{2\pi\tau}{T} C(z) - \sin \frac{2\pi\tau}{T} S(z)\right] + O(\rho - \rho_c) ,$$

where τ is fixed in the range $0 \le \tau < T$. As ρ is increased towards ρ_c in the presence of finite perturbations the plate will be seen to oscillate. The oscillations themselves will have shape mainly determined by the base profile and the function $C(z)$, with a slight amount of "sloshing" due to $S(z)$. For small amplitude perturbations, the oscillations will decay. For larger amplitude perturbations, the amplitude of the oscillations will grow.

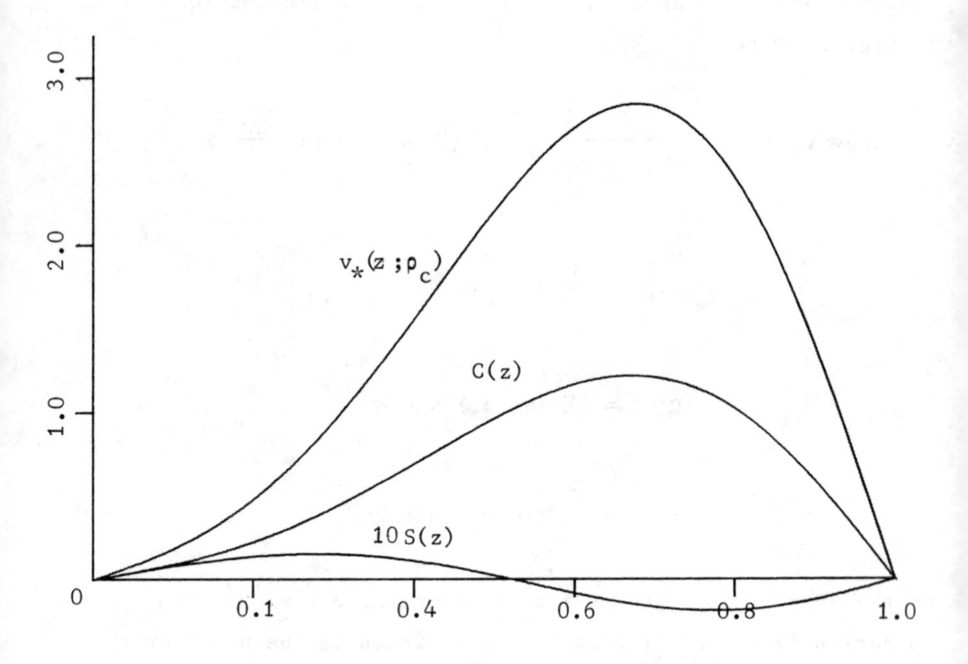

Figure 3.4. Functions $v_*(z;\rho)$, $C(z)$ and $S(z)$.

7. EXERCISES

Exercise 1.

Derive the algorithm outlined in Section 3 of this Chapter

for the evaluation of μ_2, τ_2 and β_2 from the results of Chapter 1. Take $q = v_1/2$, $q^* = 2\bar{u}_1$. Convince yourself that this algorithm produces the same results as the Recipe-Summary of Chapter 2.

Exercise 2.

Write a FORTRAN program which uses BIFOR2 to analyze the Hopf bifurcation in Langford's system (Chapter 4, Section 2). See Appendices E and F for guidelines in writing the program. Compare your program with subroutines LN and LNFUN on the microfiche.

Exercise 3.

Kubiček [73] considers the dynamic behaviour of two stirred tank reactors with first order reaction and recycle, described by the system

$$\dot{x}_1 = (1 - \Lambda)x_3 - x_1 + \alpha(1 - x_1)\exp_\gamma(x_2)$$

$$\dot{x}_2 = (1 - \Lambda)x_4 - x_2 + \alpha B(1 - x_1)\exp_\gamma(x_2) - \beta_1(x_2 - T_c)$$

$$\dot{x}_3 = (\alpha/\beta)[x_1 - x_3 + \beta(1 - x_3)\exp_\gamma(x_4)]$$

$$\dot{x}_4 = (\alpha/\beta)[x_2 - x_4 + \beta B(1 - x_3)\exp_\gamma(x_4) - \beta_2(x_4 - T_c)]$$

in which $\exp_\gamma(x) = \exp(x/(1 + x/\gamma))$, α is taken as the bifurcation parameter and the remaining parameters have values $\gamma = 1000$, $B = 12$, $\beta_1 = \beta_2 = 2$, $T_c = 0$, $\Lambda = .8$ and $\beta = .2$. Hopf bifurcations are found to occur at $\alpha = \alpha_1, \alpha_2$ and α_3 where $\alpha_1 = .09556$, $\alpha_2 = .1574$ and $\alpha_3 = .2730$. Write a program which uses BIFOR2 to analyze these bifurcations. Show in particular, that $\mu_2(\alpha_1) < 0$, $\mu_2(\alpha_2) < 0$ and $\mu_2(\alpha_3) < 0$. Discuss the claims in [73] concerning periodic solutions of the system.

Exercise 4.

Construct (or obtain) a code which will integrate the general initial value problem

$$y' = f(x,y), \quad y(x_0) = y_0, \quad y \in \mathbb{R}^n ,$$

from $x = x_0$ until the first $x > x_0$ such that $g(x,y(x)) = 0$, where $g(x,y)$ is a user-supplied real valued function. Check the code integrating the system $y_1' = y_2$, $y_2' = -y_1$ with $y_1(0) = 0$, $y_2(0) = 1$ until the first $x > 0$ such that $y_1(x) = 0$.

Exercise 5.

a) Using the code from Exercise 4, integrate the initial value problem

$$y_1' = y_1(y_2 - 1/3) , \quad y_1(0) = r_0$$

$$y_2' = (2y_2/3) - (y_1^2 + y_2^2) , \quad y_2(0) = 1/3$$

until the second positive x such that $y_2(x) = 1/3$. Here, $r_0 = .2$. Sketch the trajectory in the (y_1,y_2) plane.

b) Repeat part a), but for $r_0 = .275$. (These trajectories are cross sections of the tori, Figure 2.1).

Exercise 6.

A scheme sometimes used to compute unstable periodic solutions in $\mathbb{R}^2$ consists of integrating backwards in time, starting from an estimate y_0 of a point on the desired periodic orbit.

a) Under what conditions will this scheme succeed?

b) Suppose the system is in $\mathbb{R}^3$. Under what conditions will this scheme succeed?

Exercise 7.

Construct (or obtain) a code which will solve the general nonlinear algebraic system $f(y) = 0$, for $y = y_* \in \mathbb{R}^n$. The code must be derivative-free, that is, it must require only function evaluations. (It may, however, approximate the Jacobian by numerical differencing). Check the code by solving for the stationary points of the ordinary differential system in Exercise 5.

Exercise 8.

Consider the steam engine-governor system (6.4). Suppose κ and ρ are fixed. Let a function $\varphi(\xi,\varepsilon) = (\varphi_1(\xi_1,\xi_2;\varepsilon), \varphi_2(\xi_1,\xi_2;\varepsilon))^T$ be defined as follows.

Let $\gamma = \xi_1$, then integrate the initial value problem consisting of (6.4) with initial conditions $x_1(0) = \varepsilon + \cos^{-1}\rho$, $x_2(0) = 0$, $x_3(0) = \xi_2$ until $t = t_2$, the second time $t > 0$ such that $x_2(t) = 0$. Define $\varphi_1 = x_1(t_2) - x_1(0)$, $\varphi_2 = x_3(t_2) - x_3(0)$.

a) Write a program to evaluate $\varphi(\xi;\varepsilon)$. Use the code from Exercise 4 to solve the initial value problems.

b) For $\kappa = .1$, $\rho = .2$ and $\varepsilon = .1$, solve the nonlinear system $\varphi(\xi;\varepsilon) = 0$ for $\xi = \xi^*(\varepsilon)$. Take $\xi = (\gamma_c,0) = (2\kappa\rho^{3/2},0)$ as the initial estimate. Use the code from Exercise 7.

c) Repeat part b) for $\varepsilon = .4, .2, .1, .05$ and $.025$. For each of these values of ε, compute $(\xi_1^*(\varepsilon) - \gamma_c)/\varepsilon^2$ and compare with $\mu_2(\kappa,\rho)$ from Table 3.2.

d) Estimate how much computer time would be needed to evaluate every entry of Table 3.2, using the method of part c).

e) Repeat part c) for $(\kappa,\rho) = (.3,.7)$, $(.3,.8)$, and $(.3,.781)$.

Remark. In this last case, for values of $\gamma < \gamma_c$, $\gamma \approx \gamma_c$, there are two small amplitude periodic solutions in addition to the equilibrium solution. Such multiple periodic solutions often arise near curves (surfaces, hypersurfaces) $\mu_2 = 0$ in the space of parameters which remain after the bifurcation parameter has been selected. The explanation is: near such a curve (surface or hypersurface), it often happens that $0 < |\mu_2| \ll |\mu_4|$, so there can be two small positive solutions ε to

$$\nu - \nu_c = \mu(\varepsilon) = \mu_2\varepsilon^2 + \mu_4\varepsilon^4 + O(\varepsilon^6),$$

provided $\nu \approx \nu_c$ and $\operatorname{sgn}(\nu - \nu_c) = \operatorname{sgn} \mu_2 = -\operatorname{sgn} \mu_4$.

Exercise 9.

Suppose that the autonomous system $\dot{x} = f(x)$, $x \in \mathbb{R}^n$, f smooth, has a nontrivial periodic orbit $x = p(t)$, $-\infty < t < \infty$. Let x^0 denote an estimate of a point on the periodic orbit, and let T_0 denote an estimate of the period T . Consider the Poincaré map ψ defined as follows.

Let $u_n = f(x^0)/|f(x^0)|$, let $u_1, \ldots, u_{n-1}$ denote unit vectors chosen so that $u_i \cdot u_j = \delta_{ij}$ $(i,j = 1,\ldots,n)$, and let Π_0 denote the hyperplane containing x^0 and having normal vector u_n . For $y \in \mathbb{R}^{n-1}$, $|y|$ sufficiently small, integrate the initial value problem $\dot{x} = f(x)$, $x(0) = x^0 + \Sigma_1^{n-1} y_j u_j$ until the time $t = \tau$ closest to T_0 such that $x(\tau) \in \Pi_0$. Define ψ by means of $\psi_j = u_j \cdot (x(\tau) - x^0)$, $(1 \le j \le n - 1)$.

a) Under the assumptions that i) $p(t)$ is linearly stable, and ii) x^0 and T_0 are sufficiently 'good', show that α) the iterates $y^0 = 0$, $y^{k+1} = \psi(y^k)$, $(k = 0,1,2,\ldots)$ of the Poincaré map are all well defined, β) the sequence $y^0, y^1, y^2, \ldots$ converges to a point y^* which corresponds to a point on the periodic orbit, γ) $|y^* - y^k| < C_1 r^k$, $(k = 0,1,2,\ldots)$ for some constants $C_1 > 0$, $0 < r < 1$.

b) A scheme sometimes used to compute orbitally asymptotically stable periodic solutions consists of integrating the initial value problem $\dot{x} = f(x)$, $x(0) = x_0$ forwards in t for as long as it takes for the computed solution to 'resemble' a periodic solution. The scheme amounts to iterating the Poincaré map. Explain why this scheme tends to work poorly when computing small amplitude periodic solutions which arise by the mechanism of Hopf bifurcation.

c) Another scheme sometimes used to compute periodic solutions consists of solving the system of algebraic equations $\varphi(y) \equiv \psi(y) - y = 0$. Under the assumptions that i) only a

single characteristic exponent associated with $p(t)$ has zero real part, ii) x^0 and T_0 are sufficiently 'good', show that α) the Newton iterates $y^0 = 0$, $y^{k+1} = y^k - [\partial\varphi/\partial y(y^k)]^{-1}\varphi(y^k)$, $(k = 0,1,2,\ldots)$ are all well defined, β) the sequence $y^0, y^1, y^2, \ldots$ converges to a point y^* which corresponds to a point on the periodic orbit, and γ) $|y^* - y^{k+1}| < C_2|y^* - y^k|^2$, $(k = 0,1,2,\ldots)$ for some constant $C_2 > 0$.
(In practice, it is more convenient to evaluate the Jacobian $\partial\varphi/\partial y$ by numerical differencing than by integrating the variational equations. Modifications of Newton's method, designed to minimize the number of Jacobian evaluations, are also appropriate.)

d) Suppose that the system $\dot{x} = f(x;\nu)$, $x \in R^n$, f smooth, exhibits a Hopf bifurcation at $\nu = \nu_c$, and let $p = p_\varepsilon(t)$ denote the periodic solution which arises for $\nu - \nu_c = \mu(\varepsilon) = \mu_2\varepsilon^2 + O(\varepsilon^4)$, $0 < \varepsilon < \varepsilon_0$. Suppose that $\mu_2 \neq 0$. We conjecture that, for all sufficiently small ε, the estimates $x^0 = x_*(\nu_c) + \varepsilon \mathrm{Re}\, v_1$, $T_0 = 2\pi/\omega(\nu_c)$ are good enough to guarantee convergence of the simple shooting scheme in part c). We further conjecture that if $\beta_2 < 0$, the scheme in part b) will converge with the same estimates. Investigate these conjectures.

Exercise 10.

Suppose that the function $\varphi(y) = \psi(y) - y$ from Exercise 9 is actually computed as $\varphi(y) + e(y)$, where $e(y)$ includes the effects of accumulated truncation and roundoff error during solution of the initial value problem.

a) Let E denote an estimate of $|e(y)|$. Explain the choice $\delta = (|x^0|E)^{\frac{1}{2}}$ of increment for use in the one-sided difference approximation

$$\partial\varphi/\partial y(y)]_{ij} \approx (\varphi_i(y + \delta\hat{e}_j) - \varphi_i(y))/\delta \,, \quad (i,j = 1,\ldots,n-1) \,.$$

(See Appendix D.).

b) Let $\varkappa$ denote the condition number of the Jacobian matrix $\partial\varphi(y^*)/\partial y$. Explain why the simple shooting scheme in 9c) with Jacobian evaluations by one-sided differencing as above, may fail if $\varkappa$ is comparable with or exceeds $(|x^0|/E)^{\frac{1}{2}}$.

c) Discuss how the condition number is affected by the period and characteristic exponents.

d) How is the condition in part b) changed if two-sided differencing with $\delta = (|x^0|E)^{1/3}$ is used instead of one-sided differencing ?

e) How is the condition in part b) changed if the Jacobian $\partial\varphi/\partial y$ is evaluated by numerical integration of the variational equations ? (Assume that this Jacobian is actually computed as $\partial\varphi/\partial y + \mathcal{E}(y)$, where $\|\mathcal{E}(y)\|$ may be estimated by E.)

Remark. This exercise explores a limitation of the technique of simple shooting, and offers an explanation of the numerical difficulties encountered in [44] . For schemes which overcome this particular difficulty, see 'Codes for Boundary-Value Problems in Ordinary Differential Equations', edited by B. Childs, M. Scott, J. W. Daniel, E. Denman and P. Nelson, Lecture Notes in Computer Science, vol. 76, Springer-Verlag N. Y. (1979). See in particular the paper, An analysis of the stabilized march, by M. R. Osborne, the originator of multiple shooting.

CHAPTER 4. APPLICATIONS: DIFFERENTIAL-DIFFERENCE AND INTEGRO-DIFFERENTIAL EQUATIONS (BY HAND)

1. INTRODUCTION

Many models of biological phenomena involve delays between interactions. For example, the Hutchinson-Wright equation [111]

$$\frac{dx(t)}{dt} = -a\ x(t-1)[1+x(t)] \qquad (x \in \mathbb{R}^1) \tag{1.1}$$

was introduced by Hutchinson [58] to describe the growth of a single species. The constant time delay simulates reproductive or response lag; for example, for a species of fish this represents a lag between spawning and reproductive maturity of the offspring (Walter [107]). In this Chapter we shall show how the theory of Chapter 1 can be extended to include classes of delay differential equations and integro-differential equations, which may incorporate several distinct, constant delays or a continuous spectrum of delays. The Center Manifold Theorem will again be used (see Appendix A) to reduce the equations we consider to a single ordinary differential equation in one complex variable on a center manifold. This reduction to the two-dimensional case is more dramatic than that effected in Chapter 1 because of the infinite dimensional character of the functional differential equations we consider. Once the problem has been reduced to a two-dimensional one, it is then possible to use the formulas of Chapter 1 to determine the stability, direction of bifurcation, periods, and asymptotic forms of the small amplitude periodic solutions bifurcating from the steady state.

We shall apply this theory to several examples, including the Hutchinson-Wright equation, a two-delay model of Stirzaker [103] for the population of the field role (Microtus agrestis), and a vegatation-herbivore-carnivore model introduced by May [84]. Much of the material in this Chapter is taken from Kazarinoff, vandenDriessche, and Wan [71].

2. THEORY AND ALGORITHM FOR DIFFERENTIAL-DIFFERENCE EQUATIONS

Let $C^k[-r,0]$ denote the space of real, n-dimensional vector valued functions on the interval $[-r,0]$ all of whose components have k continuous derivatives. When $k = 0$, the superscript will be omitted.

Consider an autonomous system

$$\frac{dx(t)}{dt} = L_\mu x_t + f(x_t(\cdot),\mu) \qquad (t > 0,\ \mu \in \mathbb{R}^1) , \tag{2.1}$$

where for some $r > 0$

$$x_t(\theta) = x(t+\theta), \quad x:[-r,0] \to \mathbb{R}^n , \quad \theta \in [-r,0] .$$

Here L_μ is a one-parameter family of continuous (bounded) linear operators

$$L_\mu : C[-r,0] \to \mathbb{R}^n .$$

The operator

$$f(\cdot,\mu) : C[-r,0] \to \mathbb{R}^n$$

contains the nonlinear terms, beginning with at least quadratic terms

$$f(0,\mu) = 0 \ , \quad D_x f(0,\mu) = 0 \ .$$

The equation (1.1) is an example equation of the form (2.1).) or simplicity, and because it is true in the examples we onsider, we assume that $f(\cdot,\mu)$ is analytic and that f and μ depend analytically on the bifurcation parameter μ for $|\mu|$ small. Actually, in applications where computation of $_2$, τ_2, β_2 is sufficient a C^4 hypothesis on L will do. One onsiders solutions of (2.1) to be elements of

$$C : C[-r,0] \to \mathbb{R}^n \ ,$$

hat is, solutions map continuous initial data into $\mathbb{R}^n$. An rbit corresponding to a solution x of (2.1) is a curve in C raced out by the family of functions $x_t(\cdot)$ $(x_t(\theta) = x(t+\theta))$ s t ranges over $(0,\infty)$; the orbit of a periodic solution is closed curve in C .

For the basic existence and uniqueness theory for smooth olutions of the initial value problem for (2.1) we refer the eader to books by Hale [40] and Halanay [38], and an article y Chafee [16]. The theory we develop below depends upon the xistence of a center manifold for the suspended system for (2.1), hat is, the system consisting of (2.1) together with the extra quation $\dot{\mu} = 0$. Chafee [16] has proved that there exists such center manifold $\mathbf{C}$ under the hypotheses we assume, as has laeysson [19]. As in Chapter 1, the individual periodic orbits ill belong to slices $\mathbf{C}_\mu$ (μ = constant) of $\mathbf{C}$.

Chafee's theory must be modified in the case of integro-ifferential equations over an infinite interval in θ . We iscuss this further in Section 5. Integral equations over inite intervals $[-r,0]$ are included in the class of equations epresented by (2.1).

Our immediate objective is to cast (2.1) into the form

$$\dot{x}_t = A(\mu)x_t + Rx_t ,$$

which is a more mathematically pleasing one because this equation involves a single unknown vector x_t rather than both x and x_t. We begin by transforming the linear problem $\dot{x} = L_\mu x_t$. By the Riesz representation theorem, there exists an $n \times n$ matrix-valued function

$$\eta(\cdot,\mu) : [-r,0] \to \mathbb{R}^{n^2}$$

whose components each have bounded variation and such that for all $\phi \in C[-r,0]$,

$$L_\mu \phi = \int_{-r}^{0} d\eta(\theta,\mu)\phi(\theta) .$$

In particular,

$$L_\mu x_t = \int_{-r}^{0} d\eta(\theta,\mu)x(t+\theta) .$$

For example, if

$$L_\mu x_t = -(\frac{\pi}{2}+\mu)\, x(t-1) , \tag{2.2}$$

then

$$d\eta(\theta,\mu) = -(\frac{\pi}{2}+\mu)\delta(\theta+1) ,$$

where $\delta(\theta)$ is the Dirac delta function. The choice $d\eta(\theta,\mu) = \mu\delta(\theta)$ corresponds to the ordinary differential equation $\dot{x} = \mu x$.

We make the usual Hopf assumptions on the spectrum

$$\sigma(\mu) = \{\lambda | \det(\lambda I - L_\mu e^{\lambda\theta} I) = 0\}$$

of L_μ ; namely,

(1) there exists a pair of complex, simple eigenvalues $\lambda(\mu)$ and $\bar{\lambda}(\mu)$ such that

$$\lambda(\mu) = \alpha(\mu) + i\omega(\mu) ,$$

where α and ω are real and

$$\alpha(0) = 0 , \quad \omega(0) = \omega_0 > 0 ,$$

and

$$\alpha'(0) \neq 0 \qquad \text{(the transversality hypothesis);}$$

and

(2) all other elements of $\sigma(0)$ have negative real parts.

For example (2.2) the characteristic equation is

$$\lambda + (\tfrac{\pi}{2} + \mu)e^{-\lambda} = 0 . \tag{2.3}$$

If $\mu = 0$, $\pm i\pi/2$ are eigenvalues and all other roots of (2.3) have negative real parts at $\mu = 0$ [111]. We identify $\lambda(\mu)$ in this example by choosing $\lambda(0) = i\pi/2$. Then $\omega_0 = \pi/2$, and $\lambda'(0) = \upsilon(i + \frac{\pi}{2})$, where $\upsilon = (1 + \pi^2/4)^{-1}$. Hence $\text{Re}\,\lambda'(0) = \upsilon\pi/2 > 0$ so that the transversality condition is satisfied for scalar delay differential equations having linear part (2.2).

Next we define for $\phi \in C^1[-r,0]$

$$A(\mu)\phi = \begin{cases} d\phi/d\theta & \text{if} \quad -r \leq \theta < 0 \\ \int_{-r}^{0} d\eta(s,\mu)\phi(s) \equiv L_\mu\phi & \text{if} \quad \theta = 0 , \end{cases} \tag{2.4}$$

and

$$R\phi = \begin{cases} 0 & \text{if } -r \le \theta < 0 \\ f(\phi,\mu) & \text{if } \theta = 0 \end{cases} \quad . \tag{2.5}$$

Then since $dx_t/d\theta = dx_t/dt$, (2.1) becomes

$$\dot{x}_t = A(\mu)x_t + Rx_t \ , \tag{2.6}$$

which is an equation of the form we desired. For $\theta \in [-r,0)$, (2.6) is just the trivial equation $dx_t/dt = dx_t/d\theta$; for $\theta = 0$ it is (2.1).

We shall obtain explicit expressions for μ_2, τ_2, and β_2 only. Thus we need only to compute at $\mu = 0$, and we set $\mu = 0$ in the following. We define $q(\theta)$ to be the eigenvector for $A(0)$ corresponding to $\lambda(0)$; namely

$$A(0)q(\theta) = i\omega_0 q(\theta) \ . \tag{2.7}$$

The adjoint operator $A^*(0)$ is defined by

$$A^*(0)\alpha(s) = \begin{cases} -d\alpha(s)/ds & \text{if } 0 < s \le r \\ \int_{-r}^{0} d\eta^T(t,0)\alpha(-t) & \text{if } s = 0 \ , \end{cases} \tag{2.8}$$

where η^T denotes the transpose of η (recall that L_μ is real). We shall henceforth simply write A for $A(0)$, A^* for $A^*(0)$, and $\eta(s)$ for $\eta(s,0)$, etc. Note that the domains of A and A^* are $C^1[-r,0]$ and $C^1[0,r]$, respectively, where for convenience in computation we shall allow functions with range $\mathbb{C}^n$ instead of $\mathbb{R}^n$. Since

$$Aq(\theta) = \lambda(0)q(\theta) \ ,$$

$\bar{\lambda}(0)$ is an eigenvalue for A^* , and

$$A^*q^* = -i\omega_0 q^* \tag{2.9}$$

for some nonzero vector q^* .

To construct coordinates to describe C_0 near 0 , we need an inner product. For $\psi \in C[0,r]$ and $\phi \in C[-r,0]$ this is defined by

$$\langle \psi,\phi \rangle = \overline{\psi}(0)\cdot\phi(0) - \int_{\theta=-r}^{0} \int_{\xi=0}^{\theta} \overline{\psi}^T(\xi-\theta)d\eta(\theta)\phi(\xi)d\xi \tag{2.10}$$

Here, for a and b in C^n , $a\cdot b$ means $\Sigma_1^n a_i b_i$, where the a_i and b_i are the components of the vectors a and b . Then, as usual,

$$\langle \psi, A\phi \rangle = \langle A^*\psi, \phi \rangle$$

for $(\phi,\psi) \in D(A) \times D(A^*)$. We normalize q and q^* by the condition

$$\langle q^*,q \rangle = 1 . \tag{2.11}$$

Of course,

$$\langle q^*,\overline{q} \rangle = 0 ,$$

since $i\omega_0$ is a simple eigenvalue for A . For each $x \in D(A)$, we may then associate the pair (z,w) where $z = \langle q^*,x \rangle$ and $w = x - zq - \overline{z}\overline{q} = x - 2\,\mathrm{Re}\{zq\}$.

For x_t a solution of (2.6) at $\mu = 0$, we define

$$z(t) = \langle q^*,x_t \rangle \tag{2.12}$$

and then define

$$w(t,\theta) = x_t(\theta) - 2\,\mathrm{Re}\{z(t)q(\theta)\} . \tag{2.13}$$

On the manifold C_0, $w(t,\theta) = w(z(t),\overline{z}(t),\theta)$ where

$$w(z,\bar{z},\theta) = w_{20}(\theta)\,\frac{z^2}{2} + w_{11}(\theta)z\bar{z}$$

$$+ w_{02}(\theta)\,\frac{\bar{z}^2}{2} + w_{30}(\theta)\,\frac{z^3}{6} + \ldots \quad (2.14)$$

In effect, z and $\bar{z}$ are local coordinates for C_0 in C in the directions of q^* and $\bar{q}^*$. Note that w is real if x_t is; we shall deal with real solutions only. It is easy to see that

$$\langle q^*,w\rangle = 0 .$$

Now, for solutions $x_t \in C_0$ of (2.6),

$$\langle q^*,\dot{x}_t\rangle = \langle q^*,Ax_t + Rx_t\rangle$$

or, since $\mu = 0$,

$$\dot{z}(t) = i\omega_0 z(t) + \bar{q}^*(0)\cdot f(w(z,\bar{z},\theta) + 2\,\mathrm{Re}\{zq(\theta)\}) \quad (2.15)$$

$$= i\omega_0 z + \bar{q}^*(0)\cdot f_0(z,\bar{z})$$

which we write in abbreviated form as

$$\dot{z} = i\omega_0 z(t) + g(z,\bar{z}) . \quad (2.16)$$

Note that in (2.15), f is an operator acting upon its argument taken as a function of θ. In (2.16), however, g is a function of z and $\bar{z}$, the θ dependence having been removed.

Our next objective is to expand g in powers of z and $\bar{z}$; and then to obtain from this expansion the values of μ_2, τ_2 and β_2 using the algorithm of Chapter 1. First, it is required to derive equations for the coefficients $w_{ij}(\theta)$. We follow the procedure used in Chapter 1.

We write

$$\dot{w} = \dot{x}_t - \dot{z}q - \dot{\bar{z}}\bar{q}$$

and use (2.16) and (2.6) to obtain

$$\dot{w} = \begin{cases} Aw - 2\,\mathrm{Re}\{\overline{q}^*(0)\cdot f_0\, q(\theta)\} & \text{if } -r \le \theta < 0 , \\ Aw - 2\,\mathrm{Re}\{\overline{q}^*(0)\cdot f_0\, q(0)\} + f_0 & \text{if } \theta = 0 , \end{cases}$$

which we rewrite as

$$\dot{w} = Aw + H(z,\overline{z},\theta) \tag{2.17}$$

using (2.14), where

$$H(z,\overline{z},\theta) = H_{20}(\theta)\,\frac{z^2}{2} + H_{11}(\theta)z\overline{z} + H_{02}(\theta)\,\frac{\overline{z}^2}{2} + \dots .$$

On the other hand, on C_0 near to the origin

$$\dot{w} = w_z\dot{z} + w_{\overline{z}}\dot{\overline{z}} .$$

Using (2.14) and (2.16) to replace w_z and $\dot{z}$ and their conjugates by their power series expansions (which involve the w_{ij} !), we get a second expression for $\dot{w}$. We equate this to the right-hand side of (2.17). The result, as in Chapter 1, is an equation from which we can derive equations for the n-vectors $w_{ij}(\theta)$ $(i+j = 2,3,\dots)$. These are

$$\begin{aligned} (2i\omega_0 - A)w_{20}(\theta) &= H_{20}(\theta) \\ - Aw_{11}(\theta) &= H_{11}(\theta) \\ &\dots \end{aligned} \tag{2.18}$$

Now the H_{ij} with $i+j = 2$ do not involve any of the w_{ij}! Further, by hypothesis $2i\omega_0$ and 0 are not eigenvalues of A. Thus the first three of the equations (2.18) can be solved for w_{20}, w_{11}, and $w_{02} = \overline{w}_{20}$. At each stage the equations for w_{ij} $(i+j=k+1)$ only involve via the H_{ij} coefficients w_{ij} with $i+j \le k$. Hence the equations (2.18) can be solved successively for the w_{ij}. Only the values of w_{ij} $(i+j=2)$ are needed to compute μ_2, τ_2, and β_2. If μ_{2k}, τ_{2k}, and β_{2k} are desired for some $k > 1$, then μ must not be set equal to 0 in the previous analysis.

Once the w_{ij} are determined, the differential equation (2.16) for z can be written as

$$\dot{z} = i\omega_0 z + g_{20}\frac{z^2}{2} + g_{11}z\bar{z} + g_{02}\frac{\bar{z}^2}{2} + g_{21}\frac{z^2\bar{z}}{2} + \cdots .$$

where the coefficients g_{ij} for $i+j \le 3$ may be computed by expanding the expression

$$\bar{q}^*(0)\cdot f(zq(\theta) + \bar{z}\bar{q}(\theta) + w_{20}(\theta)\frac{z^2}{2} + w_{11}(\theta)z\bar{z} + w_{02}(\theta)\frac{\bar{z}^2}{2}) .$$

The coefficient $c_1(0)$ of the Poincaré normal form is given in terms of these g_{ij}'s by formula (5.9) of Chapter 1. Formulas (3.13), (3.21) and (4.7) of Chapter 1 then give μ_2, τ_2 and β_2.

In the examples studied below equations (2.18) have solutions

$$w_{20}(\theta) = C_1 q(\theta) + C_2\overline{q(\theta)} + Ee^{2i\omega_0\theta}$$

$$w_{11}(\theta) = C_3 q(\theta) + C_4\overline{q(\theta)} + F$$

for some C_i $(i=1,\ldots,4)$ in $\mathbb{C}$ and some vectors E and F in $\mathbb{C}^n$, where $q(\theta) = q(0)e^{i\omega_0\theta}$. Then the bifurcating periodic solutions are described by

$$x(t,\mu(\varepsilon)) = 2\varepsilon \ \mathrm{Re}[q(0)e^{i\omega_0 t}] + \varepsilon^2\,\mathrm{Re}[Ee^{2i\omega_0 t} + F] + O(\varepsilon^3)$$

for

$$0 \le t \le T(\varepsilon) = \frac{2\pi}{\omega_0}\,[1+\tau_2\varepsilon^2+\ldots] .$$

These solutions are asymptotically, orbitally stable, in the usual sense, in the center manifold. That they are also asymptotically orbitally stable in the norm of C follows from their uniqueness and the results of Chafee [16].

3. THE HUTCHINSON-WRIGHT EQUATION AND RELATED EXAMPLES

In nature the growth rate of a species' population cannot respond instantaneously to those factors which affect it but rather responds after a time lag (or lags). For example, if we assume ideal climatic conditions, because of overgrazing, the birthrate of cattle on the range depends not only upon the current population but also upon the population at past times over an interval roughly equal to the time it takes the range to regenerate. Or the delay may be a natality lag, as with salmon populations. Or both natality and other time lags may occur. In biology one of the earliest mathematical models involving time lags is due to Hutchinson [58]. He considered the population of a single species obeying a logistic growth law with a constant growth rate r modified by a maximum population factor evaluated at an earlier time:

$$\frac{dN(t)}{dt} = r[1-N(t-T)/K]\,N(t) \qquad (r,T>0) \tag{3.1}$$

Here K is the maximum population size that can be sustained. If we think of $N(t)$ as the cattle population grazing upon vegetation that takes time T to recover, then equation (3.1) says that $N(t-T) = K$ results in no growth of the population at time t.

Indeed, equation (3.1) has two equilibrium states 0 and K: 0 is linearly unstable $(r>0)$, and K is linearly stable if $rT<\pi/2$ but is linearly unstable if $rT>\pi/2$. Note that for the version of (3.1) with no delay, i.e., if

$$\frac{dN(t)}{dt} = rN(t)[1 - \frac{N(t)}{K}] , \tag{3.2}$$

then 0 is linearly unstable and K is linearly stable. Equations (3.1) and (3.2) thus provide an example of the biologists "doctrine": delays are destabilizing. At best this doctrine applies only to certain linear systems [40, p. 43]; see also [9]. For small periodic oscillations arising from nonlinear equations with delays may well be stable solutions of these equations.

We wish to study Hopf bifurcation for (3.1) with either r or T as the bifurcation parameter. After a change of variables in (3.1) only the combination rT appears. To do this we set

$$t = Ts , \quad x(s) = \frac{N(Ts)-K}{K} .$$

Then (3.1) becomes

$$\frac{dx(s)}{ds} = -(rT)\, x(s-1)[1+x(s)] ,$$

which we rewrite as

$$\dot{x}(t) = -(\frac{\pi}{2} + \mu)x(t-1)[1+x(t)] , \tag{3.3}$$

where, for convenience, the symbol t is used for the new independent variable.

Much earlier than 1948, the year Hutchinson's paper appeared, two economists, R. Frisch and H. Holme, found and studied the same equation in connection with the theory of business cycles [35]. Only in 1955 did a mathematician discover (3.3). E. M. Wright [111] studied it in connection with the distribution of prime numbers. It was proved by Jones [64] that for $\mu > 0$, (3.3) has a nontrivial periodic solution. Chow and

Mallet-Paret [18] proved existence of small amplitude periodic solutions, via the method of averaging, for $\mu > 0$. They showed that these solutions are stable and derived their approximate form:

$$x(t,\mu) = \left[\frac{40\mu}{3\pi-2}\right]^{1/2} \cos(\frac{\pi}{2} t) + O(\mu) .$$

Morris [86] used Poincaré-Lindstedt series to derive this result and an expression for the frequency:

$$\Omega(\mu) \approx \frac{\pi}{2}\left[1 - \frac{2}{5\pi}\,\frac{10\mu}{3\pi-2}\right] .$$

We confirm these bifurcation results, and we show a little more: (3.3) has $T(\varepsilon)$-periodic solutions

$$x(t,\mu(\varepsilon)) = 2\,\varepsilon\cos(\frac{\pi}{2}t) + \varepsilon^2(\frac{4}{5}\cos\pi t + \frac{2}{5}\sin\pi t) + O(\varepsilon^3)$$

if $0 \le t \le T(\varepsilon) = 4(1+\frac{2}{5\pi}\,\varepsilon^2 + O(\varepsilon^4))$, where $\varepsilon = \left[\frac{10\mu}{3\pi-2}\right]^{1/2}$.

The linearization of (3.3) about $x = 0$ is

$$\dot{x}(t) = -(\frac{\pi}{2} + \mu)\, x(t-1)$$

$$= \int_{-1}^{0} d\eta(\theta,\mu)\, x(t+\theta)$$

where $d\eta = -(\frac{\pi}{2} + \mu)\delta(\theta+1)$, δ is the Dirac δ-function. The characteristic equation is

$$\lambda + (\frac{\pi}{2} + \mu)e^{-\lambda} = 0 . \tag{3.4}$$

We first examine when this equation has pure imaginary roots $\lambda = \pm i\omega$, $\omega > 0$. If $\lambda = \pm i\omega$ with $\omega > 0$,

$$(\frac{\pi}{2} + \mu)\cos \omega = 0 \quad \text{and} \quad \omega - (\frac{\pi}{2} + \mu)\sin \omega = 0 .$$

Therefore

$$\omega = (2n+1) \frac{\pi}{2} \qquad (n = 0,1,2,\ldots) .$$

Hence

$$(2n+1) \frac{\pi}{2} - (\frac{\pi}{2} + \mu)(-1)^n = 0 .$$

Thus $\mu_n = n\pi$ $(n = 0,2,4,\ldots)$ are the critical values of μ. We shall only treat the case $n = 0$.

We have thus far shown that if $\mu = 0$, (3.4) has pure imaginary roots $\pm i\pi/2$. We next prove that if $\mu = 0$, (3.4) has no roots with positive real parts. It is obvious that (3.4) has no positive real roots if $\mu = 0$. If $\alpha + i\omega$ is a root of (3.4) at $\mu = 0$ with α and ω positive, then

$$\alpha + \frac{\pi}{2} e^{-\alpha} \cos \omega = 0 \quad \text{and} \quad \omega = \frac{\pi}{2} e^{-\alpha} \sin \omega .$$

The first of these equations implies that $\omega > \frac{\pi}{2}$, and the second implies that $\omega < \frac{\pi}{2}$. Thus at $\mu = 0$ (3.4) has no complex root with positive real and imaginary parts. Similar argument shows that at $\mu = 0$ (3.4) has no complex root with positive real part and negative imaginary part.

We have verified the hypotheses for Hopf bifurcation to occur at $\mu = 0$ except to show that

$$\operatorname{Re} \left.\frac{d\lambda}{d\mu}\right|_{\mu=0} \neq 0 .$$

Since λ is analytic in μ, we can differentiate (3.4) with respect to μ to find that

$$\left.\frac{d\lambda}{d\mu}\right|_{\mu=0} = \left.\frac{-e^{-\lambda}}{1-(\frac{\pi}{2}+\mu)e^{-\lambda}}\right|_{\mu=0} = (i+\frac{\pi}{2})\nu \ ,$$

where

$$\nu = (1+\frac{1}{4}\pi^2)^{-1} \ . \tag{3.5}$$

Hence the transversality condition is satisfied, and $\omega(0) = \pi/2$, $\alpha'(0) = \pi\nu/2$, $\omega'(0) = \nu$.

Next we choose

$$q(\theta) = e^{i\pi\theta/2} \ , \tag{3.6}$$

and hence

$$q^*(s) = D\ e^{i\pi s/2} \ , \tag{3.7}$$

where

$$D = (1+i\pi/2)\nu \ . \tag{3.8}$$

Then $\langle q^*,q\rangle = 1$ and $\langle q^*,\bar{q}\rangle = 0$. Further, in this example

$$A\phi(\theta) = \begin{cases} d\phi/d\theta & \text{if } -1 \le \theta < 0 \ , \\ -\frac{1}{2}\pi\phi(-1) & \text{if } \theta = 0 \ , \end{cases}$$

so that

$$Aq(\theta) = \frac{i\pi}{2}\, q(\theta) \qquad \text{for } -1 \le \theta \le 0 \ .$$

Also

$$R\phi(\theta) = \begin{cases} 0 & \text{if } -1 \le \theta < 0 \\ -\frac{\pi}{2}\, x(t-1)x(t) & \text{if } \theta = 0 . \end{cases}$$

If we carry out the algorithm presented in Section 2, we find that a straightforward but tedious computation yields:

$$H_{11} = 0 , \quad w_{11} = 0 ,$$

$$\frac{1}{2} H_{20} = \begin{cases} \frac{-i\pi}{2}[\bar{D}q(\theta) + D\bar{q}(\theta)] & \text{if } -1 \le \theta < 0 , \\ -i\pi\nu + i\pi/2 & \text{if } \theta = 0 , \end{cases}$$

where q and D are defined by (3.6) and (3.8); and

$$w_{20} = C_1 q(\theta) + C_2 \bar{q}(\theta) + Ee^{i\pi\theta} ,$$

where

$$C_1 = -2\bar{D} , \quad C_2 = -\frac{2}{3} D , \quad \text{and} \quad E = 2(2-i)/5 .$$

Consequently,

$$g_{20} = \pi i\bar{D} , \quad g_{11} = 0 , \quad g_{02} = -\pi i\bar{D} ,$$

and

$$\frac{g_{21}}{2} = -\frac{\pi\bar{D}}{2}\,[iC_2 + \frac{(i-1)}{2}\, E] .$$

It easily follows that

$$\mu_2 = \frac{3\pi-2}{10} > 0 \tag{3.9}$$

(bifurcation takes place for $\mu > 0$) ,

$$\beta_2 = -\frac{\nu\pi}{5}\left(\frac{3\pi}{2} - 1\right) < 0 \tag{3.10}$$

(the bifurcating periodic solutions are asymptotically, orbitally stable), and

$$\tau_2 = \frac{2}{5\pi} \; . \tag{3.11}$$

Thus

$$T(\mu(\epsilon)) = 4\left(1 + \frac{4\mu}{\pi(3\pi-2)} + O(\mu^2)\right) \; ; \tag{3.12}$$

and

$$x(t,\mu(\epsilon)) = \left[\frac{40\mu}{3\pi-2}\right]^{1/2} \cos\left(\frac{\pi}{2}t\right) + \frac{10\mu}{3\pi-2}\left(\frac{4}{5}\cos\pi t + \frac{2}{5}\sin\pi t\right) + O(\mu^{3/2}) \; , \tag{3.13}$$

with $\mu = \mu_2\epsilon^2 > 0$ if $0 \leq t \leq T(\mu(\epsilon))$. We have confirmed these results by a numerical computation of the solution that is independent of the Hopf theory.

The class of examples

$$\dot{x}(t) = -\left(\frac{\pi}{2} + \mu\right)x(t-1) + \sum_{i+j=3} s_{ij}x^i(t)x^j(t-1) \tag{3.14}$$

$$(x \in R^1 \; , \; \mu \text{ and } s_{ij} \text{ real parameters})$$

involving cubic nonlinearities, but with the same linear part as in the Hutchinson-Wright equation, is easily treated by our theory. For special choices of the s_{ij} several mathematicians have studied (3.14), namely Grafton [37], Jones [65], Hausrath [47], and Stirzaker [103]. Since the linear part of (3.14) is identical to that of (3.3), we need not repeat the linear analysis at the beginning of this Section; but we can go directly to the restriction of (3.14) to the center manifold $\mathbb{C}$. At $\mu = 0$ this restriction is

$$\dot{z}(t) = \frac{i\pi}{2} z(t) + \bar{D} \sum_{i+j=3} s_{ij}[w(z,\bar{z},0)$$

$$+ 2 \operatorname{Re}\{z(t)q(0)\}]^i [w(z,\bar{z},-1)$$

$$+ 2 \operatorname{Re}\{z(t)q(-1)\}]^j \ . \qquad (3.15)$$

Since the nonlinearities in (3.15) are cubic, the vectors $w(z,\bar{z},0)$ and $w(z,\bar{z},-1)$ do not contribute to $c_1(0) = g_{21}/2$; and we easily compute that

$$\mu_2 = -\frac{2}{\pi}[3s_{30} + s_{12} - (\frac{\pi}{2}) s_{21} - (\frac{3\pi}{2}) s_{03}] ,$$

$$\beta_2 = -\pi\nu\mu_2 , \qquad (3.16)$$

$$\tau_2 = \frac{4}{\pi^2}[s_{12} + 3s_{30}] .$$

Thus the periodic solutions bifurcating from equilibrium have the representation

$$x(t,\mu(\epsilon)) = 2\epsilon \cos(\frac{\pi}{2} t) + O(\epsilon^3) \qquad (3.17)$$

for $0 \leq t \leq 4[1+\tau_2\epsilon^2+O(\epsilon^4)]$, where $\epsilon = (\mu/\mu_2)^{1/2}$.

Another interesting class of examples involving a single delay derives from models of haematopoiesis (blood cell production) introduced by Mackey and Glass [79]. In the bone marrow (in man's, for example) there are pluripotential stem cells that according to poorly understood control mechanisms [79] evolve to become constituents of the blood, e.g., erythrocytes (red blood cells). When the oxygen level of the blood falls below a certain critical level, a control substance, erythropoietin, carries instructions to the pluripotential stem cells to increase the production of erythrocytes. Control substances for other blood constituents are conjectured to exist as well. Mackey and Glass's models have the form

$$\frac{dx(t)}{dt} = -ax(t) + \Lambda(x_t(\cdot)), \tag{3.18}$$

where $x(t)$ is the cell density at time t, a is a constant cell death rate, and Λ is a smooth operator describing the generation of cells of a particular kind, e.g., erythrocytes. A particular choice of Λ made by Mackey and Glass is

$$\Lambda(x_t(\cdot)) = \frac{\beta_0 \theta^n x(t-\tau)}{\theta^n + x^n(t-\tau)} . \tag{3.19}$$

This models production of cells as a single-humped function of the cell density at a lagged time $t - \tau$. In (3.19) β_0, θ, and n are positive constants adjusted to fit experimental data. We shall assume that $\beta_0 > a > 0$, $nB > 2$ and $6aB > \beta_0$, where

$$B = \frac{\beta_0 - a}{\beta_0} .$$

This is consistent with the data. For the system (3.18) - (3.19) there is a unique steady state $x_e = \theta[B\beta_0/a]^{1/n}$. Linearized about this equilibrium (3.18) - (3.19) is

$$\frac{du(t)}{dt} = -au(t) - bu(t-\tau) \quad (u = x - x_e,\ b = \gamma(nB-1)) . \tag{3.20}$$

The characteristic equation for (3.20) is $\lambda + a + be^{-\lambda\tau} = 0$. Let τ be the bifurcation parameter. This analysis continues as Exercises 1 - 3 at the end of this Chapter.

4. AN EXAMPLE WITH TWO DISTINCT LAGS

The field vole (Microtus agrestis) is a small grass-eating rodent. It is a stable food for hawks, owls, crows, weasels, stoats, foxes, etc. In simple habitats such as plantations and open grasslands the concentration $x(t)$ of field voles is oscillatory with peaks at four-year intervals. In mixed habitats less periodicity is observed. Stirzaker [103] has argued that there are two principal contributions to the growth rate $\dot{x}(t)$. The field vole mates after the age of seven weeks. It has a life expectancy of 60-70 weeks. But the reproductive productivity of voles in a simple habitat is affected by crowding and thus will be a complicated average of the population at previous times.

Voles are easy prey for their predators. Hence one can assume that the consumption of voles by predators depends only on the capacity of the predators and their number, i.e., is some linear function of $p(t)$, the predators' concentrations at time t. But the number of mature predators depends on the number of infants reaching maturity in previous generations and involves additional time lags, as well as the availability of prey to feed the young. Hence $p(t)$ is a complicated function of the vole population at previous times.

Thus $\dot{x}(t)$ is a complicated function of x evaluated at several (or even infinitely many) previous times. Of course, the above discussion is an oversimplified one. But it gives an explanation for the presence of two or more delays in even a simple model of vole populations in homogeneous habitats.

Having given this discussion, we leave it to the reader to study Stirzaker's paper if she or he desires to learn more about field vole populations; and instead we study an idealized model equation having nothing to do with voles but which has been studied previously by different methods. This model quation is

$$\dot{x}(t) = -\left(\frac{\pi\sqrt{3}}{9} + \mu\right)[x(t-1) + x(t-2)]\ (1-x^2(t)) \qquad (4.1)$$

$$(x \in R^1)\ .$$

aplan and Yorke [68] proved that for $\mu > 0$, (4.1) has eriodic solution of period 6. Jones [65] also proved xistence of periodic solutions of (4.1) for $\mu > 0$, and he omputed them numerically, obtaining a period of 6. We shall ompute μ_2, τ_2 and β_2 for (4.1) and give the form of the eriodic solutions. We shall find that $\tau_2 = 0$ so that the eriod $T(\mu) = 6(1+0(\mu^2))$.

Linearized about $x = 0$ equation (4.2) has the solution

$$q(t) = e^{i\pi t/3} \qquad (4.2)$$

t $\mu = 0$. We identify the λ of Section 2 by choosing λ to e that characteristic value with $\lambda(0) = i\pi/3$. The haracteristic equation for

$$\dot{x}(t) = -\left(\frac{\pi\sqrt{3}}{9} + \mu\right)\ [x(t-1) + x(t-2)] \qquad (4.3)$$

s

$$\lambda = -(\delta+\mu)(e^{-\lambda}+e^{-2\lambda}) \qquad \left(\delta = \frac{\pi\sqrt{3}}{9}\right) . \qquad (4.4)$$

e shall prove that at $\mu = 0$ this equation has exactly two ure imaginary roots $\pm i\pi/3$ and no root with positive real art. The straight line with equation $u = -v/\delta$ does not ntersect the curve $u = e^{-v} + e^{-2v}$ for $v \geq 0$. Thus, at $= 0$, (4.4) has no nonnegative real roots.

For (4.4) to have a pure imaginary root $\lambda = i\omega$ at $\mu = 0$ t must be that

$$\cos\omega + \cos 2\omega \equiv 2\cos\frac{\omega}{2}\cos\frac{3\omega}{2} = 0\ , \tag{4.5}$$

and

$$\omega = \delta(\sin\omega + \sin 2\omega)\ . \tag{4.6}$$

If $\cos\omega/2 = 0$, the second equation is not satisfied. Thus the pairs $(\pm\omega_n, \delta_n)$, where

$$\omega_n = \left(\frac{2n+1}{3}\right)\pi \qquad (n = 0,1,2,\ldots, \text{ and } n \not\equiv 1 \bmod 3)$$

and

$$\delta_n = \frac{\omega_n}{2\sin\omega_n}$$

yield all solutions of the simultaneous equations (4.5) and (4.6). But $\delta_n > \delta = \pi\sqrt{3}/9$ if $n > 0$. Hence $\pm i\pi/3$ are the only pure imaginary roots of (4.4) at $\mu = 0$.

It remains to prove that (4.4) has no complex root with positive real part for $\mu = 0$. Suppose $\lambda = \alpha + i\omega$, $\alpha > 0$. Then λ is a root of (4.4) at $\mu = 0$ if and only if

$$\alpha = -\delta(e^{-\alpha}\cos\omega + e^{-2\alpha}\cos 2\omega) \tag{4.7}$$

and

$$\omega = \delta(e^{-\alpha}\sin\omega + e^{-2\alpha}\sin 2\omega)\ . \tag{4.8}$$

If $\alpha > 0$, (4.8) implies that $|\omega| \le 2\delta$. Therefore we only need to look for solutions of (4.7) - (4.8) for $\omega \in (0, 2\delta]$. Now (4.7) implies

$$-\frac{\alpha}{\delta\cos\omega} = e^{-\alpha} + \frac{\cos 2\omega}{\cos\omega}e^{-2\alpha} \tag{4.9}$$

Since

$$\frac{\cos 2t}{\cos t} \geq -1$$

if $t \in (0, \frac{\pi}{3}]$, the right-hand side of (4.9) is positive there. Hence we need only look for solutions of (4.7) - (4.8) for

$$\omega \in (\frac{\pi}{3}, 2\delta] .$$

But

$$G(\omega) = \omega - \delta e^{-\alpha} \sin \omega - \delta e^{-2\alpha} \sin 2\omega$$

has a positive derivative for such ω and $G(\frac{\pi}{3}) > 0$. Hence (4.4) has no roots with positive real parts for $\mu = 0$.

Finally, the characteristic equation (4.4) may be differential with respect to μ to yield

$$\lambda'(0) = i\upsilon\sqrt{3}\,[1 + \delta(\frac{1}{2} - \frac{3i\sqrt{3}}{2})] ,$$

where

$$\upsilon^{-1} = \frac{\pi^2}{4} + (1 + \frac{\delta}{2})^2$$

Hence

$$\omega(0) = \frac{\pi}{3} , \ \alpha'(0) = \pi\upsilon\sqrt{3/2} > 0 , \ \text{and} \ \omega'(0) = \upsilon\frac{\sqrt{3}}{2}(1+\delta) .$$

In particular, the transversality condition is satisfied, and we may apply the Hopf bifurcation theory to (4.1) at $\mu = 0$.

Recall that q is defined by (4.2). We define

$$q^*(\theta) = De^{i\pi\theta/3} ,$$

where

$$D = \upsilon[1 + \frac{\delta+i\pi}{2}] . \tag{4.10}$$

Then if

$$\langle \phi,\psi \rangle = \bar{\phi}(0)\psi(0) - \int_{\theta=-2}^{0} \int_{\xi=0}^{\theta} \bar{\phi}(\xi-\theta)\psi(\xi)d\xi d\eta(\theta) ,$$

where

$$d\eta(\theta) = -\delta\ [\delta(\theta+2) + \delta(\theta+1)] ,$$

$$\langle q^*,q \rangle = 1 \quad \text{and} \quad \langle q^*,\bar{q} \rangle = 0 .$$

Also

$$A\phi(\theta) = \begin{cases} \dfrac{d\phi}{d\theta} & \text{if } -2 \le \theta < 0 \\ -\delta\ [\phi(-1) + \phi(-2)] & \text{if } \theta = 0 , \end{cases}$$

and

$$R\phi(\theta) = \begin{cases} 0 & \text{if } -2 \le \theta < 0 , \\ \delta[\phi(-1) + \phi(-2)]\phi^2(0) & \text{if } \theta = 0 . \end{cases}$$

Thus with z and w defined as in Section 2

$$\dot{z}(t) = \frac{i\pi}{3} z + \bar{q}^*(0)f_0 , \tag{4.11}$$

where

$$\bar{q}^*(0)f_0 = \bar{D}\delta\ \left(\sum_{j=1}^{2} [w(-j) + 2\ \mathrm{Re}\{z(t)q(-j)\}]\right) \times$$

$$\times \left(w(0) + 2\ \mathrm{Re}\{z(t)q(0)\}\right)^2 ,$$

with $w(-j) = w(z,\bar{z},-j)$. Since the nonlinearity in (4.11) is cubic, $c_1(0) = g_{21}/2$. It is a straightforward computation to obtain g_{21} from (4.11):

$$g_{21} = -\frac{2i}{3}\pi\bar{D},$$

where D is given by (4.10). Thus

$$\mu_2 = \frac{\pi\sqrt{3}}{9} > 0$$

(bifurcation occurs for $\mu > 0$),

$$\beta_2 = -\frac{\pi^2 \upsilon}{3} < 0$$

(the bifurcating periodic solutions are asymptotically, orbitally stable),

$$\tau_2 = 0$$

$$(T(\mu(\varepsilon)) = 6[1 + O(\mu^2)] = 6[1 + O(\varepsilon^4)]), \text{ and}$$

$$x(t,\mu(\varepsilon)) = \left[2\,\frac{3\sqrt{3}\,\mu}{\pi}\right]^{1/2} \left\{\cos\frac{\pi}{3}t + O(\mu)\right\}.$$

5. UNBOUNDED DELAYS

Integro-differential equations of the form

$$\frac{dx(t)}{dt} = \int_{-\infty}^{0} x(t+s)Q(s,\mu)ds + f(x_t(\cdot),\mu)$$

$$(f(0,\mu) = Df(0,\mu) = 0), \tag{5.1}$$

where $x_t(\theta) = x(t+\theta) \quad (-\infty < \theta \le 0)$,

are not included under the theory presented in Section 2 of this Chapter. Equations with unbounded delays have been studied by several mathematicians: Lima [76], Naito [87], Stech [101] and Hale [41]. (Hale's main interest in [41] is in phenomena associated with the simultaneous variation of two or more delays (taken as bifurcation parameters) in systems of delay-differential equations. Although they are interesting, we shall not discuss these phenomena in these Notes.)

For equations of the type (5.1), initial data must be prescribed on $(-\infty,0]$ so that the effects of the "tails" of the initial data are felt by solutions for all $t > 0$. Further, to insure convergence in (5.1) an appropriate space of initial data must be introduced. We give an example of such a space. Suppose g and G are nonnegative continuous functions from $(-\infty,0]$ into $\mathbb{R}$ such that

$$g(t+s) \leq G(t)g(s) \ , \quad \int_{-\infty}^{0} g(s)ds < \infty$$

and (5.2)

$$G(\beta_0) < 1 \quad \text{for some} \quad \beta_0 \in (-\infty,0] .$$

The space

$$B_2 = \{\text{equivalence classes of measurable functions } \phi : (-\infty,0] \to \mathbb{R}^n \text{ such that } \|\phi\|_2 \equiv [|\psi(0)|^2 + \int_{-\infty}^{0} g(s)|\phi(s)|^2 ds]^{1/2} < \infty\}$$

is an appropriate space of initial data. For example, for the linear equation

$$\dot{x}(t) = r\int_{-\infty}^{0} se^{s}\, x(s+t)ds \tag{5.3}$$

we can use the space B_2 with $g(s) = e^{s/2}$ and $G(t) = e^{t/2}$.

Other possible spaces of initial data are: (a) the spaces

B_p $(p \ge 1)$ defined by replacing 2 in the definition of B_2 by p [76] and (b) the spaces

$$C_\gamma = \{\phi | \phi : (-\infty, 0] \to \mathbb{R}^n, \ \phi \text{ continuous, and}$$

$$e^{-\gamma s} \phi(s) \underset{s \to -\infty}{\longrightarrow} \text{a limit}\} ,$$

where γ is any positive number.

Infinite delays create some changes in the linear part of the analysis presented in Section 2, which we shall now redo. Consider the linear system

$$\dot{x}(t) = L_\mu x_t , \tag{5.4}$$

where $L_\mu : B_2 \to \mathbb{R}^n$ is continuous for μ in some neighborhood of 0 and $L_\mu \phi$ is linear in ϕ. Then $L_\mu \phi$ can be written as

$$L_\mu \phi = \Lambda(\mu)\phi(0) + \int_{-\infty}^{0} k(\mu,s)\phi(s)g(s)ds , \tag{5.5}$$

where the components of the n by n matrix k are L_2-functions on $(-\infty, 0)$, and $\Lambda(\mu)$ is an n by n constant matrix. (In the case of the example (5.3), $n = 1$, $\Lambda \equiv 0$, $g = e^{s/2}$, and $k = rse^{s/2}$.) The characteristic equation for (5.5) is

$$\det[\lambda I - \Lambda(\mu) - \int_{-\infty}^{0} k(\mu,s)e^{\lambda s} g(s)ds] = 0. \tag{5.6}$$

The formal adjoint equation for (5.4) is

$$\dot{y}(t) = -y(0)\Lambda(\mu) - \int_{-\infty}^{0} y(t-s)k(\mu,s)g(s)ds ;$$

where y is a row vector. Corresponding to (5.4) and its adjoint is the bilinear form

$$\langle\psi,\phi\rangle = \overline{\psi}(0)\phi(0) + \int_{-\infty}^{0}\left(\int_{s}^{0}\overline{\psi}(s-u)k(\mu,s)\phi(u)du\right)g(s)ds \qquad (5.7)$$

for $\phi \in B_2$ and ψ in some appropriate class of functions ($\psi(s)$ is an n-dimensional row vector).

With this linear setting it is possible to prove the Hopf Bifurcation Theorem for (5.1) [76]. There are now two different ways to proceed: one may show that the present situation falls within the general theory of Hopf Bifurcation [81] or one may develop the theory [101] for systems of the form (5.1). We shall sketch the second approach. Consider the properties of the spectrum of L_μ. If $x(\phi)$ is the solution of (5.4) for $\phi \in B_2$ and

$$T(t)\phi = x_t(\phi) \qquad (t \geq 0) ,$$

then $\{T(t), t \geq 0\}$ is a strongly continuous semigroup of linear operators on B_2. The condition $G(\beta_0) < 1$ in the definition of B_2 (see (5.2)) guarantees existence of a $\tau > 0$ such that the essential spectrum of $T(t)$ lies in the circle of radius $e^{-\tau t}$ [41, 42, 87]. Hence only elements of the <u>point</u> spectrum of $T(t)$ may lie outside this circle, elements determined by the characteristic equation (5.6). Therefore it is possible for the Hopf spectral hypotheses for flows to be fulfilled: for a critical value of $\mu = \mu_c$, two eigenvalues correspond to elements of $T(1)$ on the unit circle, while the other elements of the spectrum of $T(1)$ all lie inside the unit circle (cf. Chapter 5).

Further, it is possible to estimate L the radius of the essential spectrum of $T(t)$ [41]. It is a theorem that

$$r_e(T(t)) = r_e(S_0(t)) \leq |S_0(t)| ,$$

where r_e denotes the radius of the essential spectrum and where

$$S_0(t)\phi(s) = \begin{cases} 0 & \text{if } t+s \geq 0 \\ \phi(t+s) & \text{if } t+s < 0 \end{cases} \qquad (\phi \in B_2, \phi(0)=0) .$$

In the case of (5.3), if $\phi \in B_2$, $\phi(0) = 0$, then

$$|S_0(t)\phi|^2 = \int_{-\infty}^{0} |(S_0(t)\phi)(s)|^2 e^{s/2} ds ,$$

$$= \int_{-\infty}^{-t} |\phi(t+s)|^2 e^{s/2} ds ,$$

$$= e^{-t/2} \int_{-\infty}^{0} |\phi(s)|^2 e^{s/2} ds ,$$

$$= e^{-t/2} \|\phi\|_2^2 .$$

Thus for (5.3)

$$|S_0(t)| \leq e^{-t/4} .$$

It follows that in this example the essential spectrum of $T(t)$ is well behaved. In general,

$$|S_0(t)| \leq \sup_{s \in (-\infty,0)} \left(\frac{g(s-t)}{g(s)} \right)^{1/2} .$$

The elements of the discrete spectrum of $T(t)$ correspond to the eigenvalues determined by the linear operators L_μ, which in the case of the example (5.3) are easily seen to be $\pm i$ for $r = r_c = 2$. (Note that -2 belongs to the discrete spectrum of L_μ but does not have a proper eigenvector corresponding to it: $\int_{-\infty}^{0} se^s(e^{-2(s+t)})ds = \infty$.)

Lastly, we caution the reader: it takes work we have not done to apply the Center Manifold Theorem of Appendix A to (5.1).

6. A THREE-TROPHIC-LEVEL MODEL

In a mathematical description of the growth (decline) of an animal population it is more realistic to assume that the rate of growth depends on a weighted average of past population sizes than to assume that it depends on the population at a single past time, as well as the current population. (We still, however, neglect the variation with time of the age distribution within the total population.) Then dependence on (infinitely) many delays can be described by an integral. An example of such a model is described by May [84]:

$$\dot{N}(t) = rN(t)[1-\int_{-\infty}^{0} N(t+s)Q(-s)ds] - \alpha P(t)N(t)$$

$$\dot{P}(t) = -bP(t) + \beta P(t)N(t) , \tag{6.1}$$

where $N(t)$ and $P(t)$ are real scalars, r, α, b, and β are positive parameters, and

$$Q(t) = \frac{t}{T^2} e^{-t/T} \qquad (T>0) . \tag{6.2}$$

The variable N represents a herbivore population (rabbits), P represents a carnivore population (foxes), and Q introduces the effect of vegetation (lettuce). The parameter T is the expected natural recovery time of the vegetation after grazing. Hopf bifurcation theory can be applied to the system (6.1) with r chosen to be the bifurcation parameter. Various wildlife management practices affect both r and T.

To ensure existence of an equilibrium state (N^*,P^*) of (6.1) in the interior of the first quadrant in the (N,P)-plane, we assume that

$$\beta > b > 0 .$$

Then

$$(N^*,P^*)^T = \left(\frac{b}{\beta}, \frac{r}{\alpha}\left(1-\frac{b}{\beta}\right)\right)^T$$

is the unique equilibrium state of (6.1), and it lies in the first quadrant. We rewrite (6.1) about (N^*,P^*) by setting

$$N(t) = N^* + x(t), \quad P(t) = P^* + y(t).$$

The result is

$$\begin{aligned} \dot{x}(t) &= -r[N^*+x(t)]\int_{-\infty}^{0} x(t+s)Q(-s)ds - \alpha[N^*+x(t)]y(t), \\ \dot{y}(t) &= \beta x(t)[P^*+y(t)]. \end{aligned} \qquad (6.4)$$

The linearization of (6.4) about $(0,0)$ is:

$$\begin{aligned} \dot{x}(t) &= -rN^*\int_{-\infty}^{0} x(t+s)Q(-s)ds - \alpha N^* y \\ \dot{y}(t) &= \beta P^* x(t). \end{aligned} \qquad (6.5)$$

For $g(s)$ we choose $e^{s/2T}$. Then in (5.5) we make the identifications

$$\Lambda = \begin{pmatrix} 0 & -\alpha N^* \\ \beta P^* & 0 \end{pmatrix}, \quad k(r,s) = \begin{pmatrix} \frac{-rN^*}{T^2} se^{s/2T} & 0 \\ 0 & 0 \end{pmatrix}.$$

The characteristic equation for (6.5) is

$$(1+\lambda T)^2(\lambda^2+\Omega^2) + rN^*\lambda = 0, \qquad (6.6)$$

where

$$\Omega = (rb)^{1/2}(1-N^*)^{1/2} . \tag{6.7}$$

Remark. The model (6.4) is really finite dimensional. The characteristic equation gives that away. In Exercise 4 it is shown how to rewrite (6.1) as a fourth order system of ordinary differential equations. However, for general kernels Q, (6.1) is not finite dimensional, and hence we judge it fair to discuss (6.1) for the kernel (6.2) in the setting of Section 5.

For

$$r = r_c \overset{d}{=} 2[TN^* + 2bT^2(1-N^*)]^{-1} , \tag{6.8}$$

(6.6) has pure imaginary roots $\lambda(r_c) = i/T$ and $\bar{\lambda}(r_c) = -i/T$. At r_c the remaining roots of (6.6) are

$$\frac{1}{T}[-1 \pm (N^* Tr_c/2)^{1/2}] .$$

Both of these roots are negative since $0 < N^* < 1$, and so for $\phi \in B_2$ the initial value problem for (6.5) is properly posed.

It remains to verify the transversality condition. Differentiating (6.6) with respect to r, we find that

$$\lambda'(r_c) = \frac{2[N^*Tr_c + 4i]}{Tr_c[16+(N^*Tr_c)^2]} .$$

Thus

$$\omega(r_c) = \frac{1}{T} , \quad \alpha'(r_c) = \frac{2N^*}{16+(N^*Tr_c)^2} ,$$

$$\omega'(r_c) = \frac{8}{Tr_c[16+(N^*Tr_c)^2]} , \tag{6.9}$$

where $\alpha(r) = \text{Re}\,\lambda(r)$, $\omega(r) = \text{Im}\,\lambda(r)$, and the transversality condition is fulfilled.

With

$$B = -i\beta\, TP^* , \tag{6.10}$$

$$q(\theta) = \begin{pmatrix} 1 \\ B \end{pmatrix} e^{i\theta/T} \qquad (-\infty < \theta \le 0) \tag{6.11}$$

is an eigenvector of

$$A \equiv \Lambda_{r_c} + \int_{-\infty}^{0} k(r_c,s)(\cdot)g(s)ds . \tag{6.12}$$

(We shall allow, as in previous sections, functions in B_2 to be complex valued in each component. Further for $\phi \in B_2$ and ψ in its appropriate adjoint, we define

$$\langle\psi,\phi\rangle = \bar{\psi}(0)\phi(0) + r_c N^* \int_{\theta=-\infty}^{0} \int_{\xi=0}^{\theta} \bar{\psi}^1(\xi-\theta)\phi^1(\xi)Q(-\theta)d\xi d\theta . \tag{6.13}$$

Then if

$$D = \frac{2}{4-ir_c N^*T} , \qquad C = -i\alpha\, TN^* , \tag{6.14}$$

and

$$q^*(\theta) = D\begin{pmatrix} 1 \\ C \end{pmatrix} e^{i\theta/T} \qquad (0 \le \theta < \infty) , \tag{6.15}$$

we compute that

$$\langle q^*,q\rangle = 1 \quad \text{and} \quad \langle q^*,\bar{q}\rangle = 0 ,$$

as desired.

If

$$
R\phi(\theta) = \begin{cases} \begin{pmatrix} 0 \\ 0 \end{pmatrix} & \text{if } -\infty < \theta < 0 \\ \begin{pmatrix} -r_c\phi^1(0) \int_{-\infty}^{0} \phi^1(s)Q(-s)ds - \alpha\phi^1(0)\phi^2(0) \\ \beta\phi^1(0)\phi^2(0) \end{pmatrix} & \text{if } \theta = 0 , \end{cases} \tag{6.16}
$$

we can rewrite (6.4) using (6.16) and (6.12) in the form

$$\dot{x}_t = Ax_t + Rx_t .$$

Next we define $z(t)$ and $w(z,z,\theta)$ as in Section 2. Then on the centermanifold $\mathbb{C}$ for (6.4) at r_c

$$\dot{z}(t) = \frac{i}{T} z(t) + \overline{D}(f_0^1 + \overline{C}f_0^2)$$

$$= \frac{i}{T} z(t) + g(z,\overline{z}) ,$$

where

$$f_0^1 = -r_c[w^1(0)+2\text{Re } z(t)] \left[\int_{-\infty}^{0} [w^1(s)+2\text{Re}\{z(t)e^{is/T}\}]Q(-s)ds + \frac{\alpha}{r_c}[w^2(0)+2\text{Re}\{z(t)B\}] \right] ,$$

$$f_0^2 = \beta[w^1(0) + 2\text{Re } z(t)][w^2(0) + 2\text{Re}\{z(t)B\}] ,$$

in which

$$w(t) = \begin{pmatrix} w^1(t) \\ \\ w^2(t) \end{pmatrix} = w(z,\bar{z},t) .$$

The system of differential equations satisfied by w is

$$\dot{w} = Aw - 2\mathrm{Re}\{\bar{D}(f_0^1 + \bar{C}f_0^2)q(\theta)\} + \begin{cases} \begin{pmatrix} 0 \\ 0 \end{pmatrix} & \text{if } -\infty < \theta < 0 \\ \begin{pmatrix} f_0^1 \\ f_0^2 \end{pmatrix} & \text{if } \theta = 0 \end{cases} ,$$

or

$$\dot{w} = Aw + H(z,\bar{z},\theta) .$$

Hence

$$\left[\frac{2i}{T} - A\right] w_{20} = H_{20}(\theta)$$

$$= -2\bar{D}\left(\frac{ir_c}{2} - \alpha B + \beta B\bar{C}\right)q(\theta)$$

$$-2D\left(\frac{ir_c}{2} - \alpha B + \beta BC\right)\bar{q}(\theta)$$

$$\begin{cases} \begin{pmatrix} 0 \\ 0 \end{pmatrix} & \text{if } -\infty < \theta < 0 , \\ 2\begin{pmatrix} \frac{ir_c}{2} - \alpha B \\ \beta B \end{pmatrix} & \text{if } \theta = 0 , \end{cases}$$

and

$$-Aw_{11} = H_{11}(\theta) = 0 \qquad (-\infty < \theta \le 0) .$$

Solving for the w_{ij} $(i+j=2)$, we find that

$$w_{11}(\theta) = \begin{pmatrix} 0 \\ 0 \end{pmatrix}, \tag{6.17}$$

and

$$w_{20} = \bar{w}_{02} = C_1 q + C_2 \bar{q} + \begin{pmatrix} E_1 \\ E_2 \end{pmatrix} e^{2i\theta/T} , \tag{6.18}$$

where

$$C_1 = 2T\bar{D}\left(\frac{-1}{TN^*} + i\beta B\bar{C}\right) ,$$

$$C_2 = \frac{2TD}{3}\left(\frac{-1}{TN^*} + i\beta BC\right) ,$$

$$E_1 = \frac{r_c b^2 T^2 (1-N^*) + \frac{2i}{T}}{N^*\left(\frac{-3r_c N^*}{25} + i\left(\frac{3}{2T} + \frac{9r_c N^*}{100}\right)\right)} , \tag{6.19}$$

and

$$E_2 = -\frac{\beta TP^*}{2}[2\beta T + iE_1] , \tag{6.20}$$

where r_c, N^* and P^* are defined by (6.8) and (6.3) (with $r = r_c$).

The next step is to compute $g(z,\bar{z})$ and $c_1(r_c)$. We find using (6.17) - (6.18), that

$$g_{11} = 0$$

$$g_{02} = \bar{D}(-ir_c - 2\alpha\bar{B} + 2\beta\bar{B}\bar{C})$$

and

$$g_{21} = \bar{D}\left\{2i\left[\frac{-1}{TN^*} - i\beta\bar{B}\bar{C}\right]C_2 - \left[\frac{i}{TN^*} - \beta\bar{B}\bar{C} - \frac{(3+4i)}{25}r_c\right]E_1 + (\beta\bar{C}-\alpha)E_2\right\}.$$

Since $g_{11} = 0$, $\text{Re}\ c_1(r_c) = \text{Re}\ g_{21}/2$. But the quantity

$$2i\bar{D}C_2\left[\frac{-1}{TN^*} - i\beta\bar{B}\bar{C}\right]$$

is pure imaginary. Therefore

$$\text{Re}\ c_1(r_c) = \frac{1}{2}\text{Re}\left[\bar{D}\left\{-\frac{1}{2}\alpha\beta^2T^2N^*P^* + \frac{3r_c}{25} - ir_c\left(\frac{17}{50} + \frac{\beta T}{2}(1-N^*)\right)\right\}E_1 + \bar{D}(\alpha - \beta\bar{C})\beta^2T^2P^*\right]. \qquad (6.21)$$

This expression is complicated, since it involves sums of products of complex numbers. Since $\alpha'(r_c) > 0$, $\text{sgn}\ \beta_2 = -\text{sgn}\ \mu_2$; and if

$$\mu_2 = -\frac{\text{Re}(c_1(r_c))}{\alpha'(r_c)} > 0,$$

then the bifurcating periodic solutions are stable and bifurcate from equilibrium for $r > r_c$. If $\mu_2 < 0$, then these solutions are unstable and bifurcate from equilibrium for $r < r_c$.

It is not difficult to tabulate μ_2 for various parameter values. The function $\mu_2(\beta,b,T)$ is independent of α (the coefficient of the interaction term in the first of the equations (6.1)), and it possesses the symmetry expressed by

$$\mu_2(\beta,b,T) = b\mu_2(\beta/b,1,bT) .$$

This allows one to construct a table of values. (The symmetry follows from the substitutions

$$N(t) = \tilde{N}(\tilde{t}),\ P(t) = \frac{b}{\alpha}\tilde{P}(\tilde{t}),\quad t = \tilde{t}/b$$

in (6.1). The resulting system for $\tilde{N},\tilde{P}$ has the same form as (6.1) except that $\tilde{T} = bT$, $\tilde{\beta} = \beta/b$ and $\tilde{r} = r/b$ replace T, β, and r respectively, while α and β are replaced by 1 .)

Table (6.1) contains values of $\mu_2(\tilde{\beta},1,\tilde{T})$ for the values of $\tilde{\beta}$ and $\tilde{T}$ indicated. To find μ_2 for general β, b, T form $\tilde{\beta} = \beta/b$ and $\tilde{T} = bT$, and multiply the entry in the table by b. Positive entries in the table correspond to parameter combinations for which the bifurcation at r_c is supercritical and to stable periodic solutions, while negative entries would represent subcritical bifurcation to unstable periodic solutions.

In the limit $T = 0$, the model (6.1) reduces to a single Lotka-Volterra equation with no time lags for which there is no Hopf bifurcation.

If $N^*\uparrow 1$, then $P^*\downarrow 0$; and (6.1) reduces to a single equation, which can be written as

$$\dot{x}(t) = -r[1 + x(t)]\int_{-\infty}^{0} x(t + s)Q(-s)ds .$$

It is easy to evaluate μ_2 , τ_2 , and β_2 for this equation. We find that

$$N^* = 1\ ,\ r_c = \frac{2}{T},\ \lambda'(\tfrac{2}{T}) = \frac{1+2i}{10}\ ,\ \text{and}\ \ c_1(\tfrac{2}{T}) = \frac{(-1-11i/3)}{20T} .$$

Table 6.1

$\tilde{\beta} \backslash \tilde{T}$	.125	.25	.5	1	2	4	8
1.001	4.011	2.005	1.002	0.502	0.257	0.155	0.185
1.01	4.106	2.047	1.020	0.520	0.318	0.390	0.980
1.1	5.135	2.490	1.208	0.681	0.685	1.150	1.899
1.2	6.435	3.033	1.431	0.841	0.886	1.275	1.713
1.3	7.906	3.629	1.670	0.993	1.025	1.334	1.632
1.4	9.549	4.278	1.924	1.145	1.148	1.397	1.622
1.5	11.369	4.978	2.193	1.299	1.266	1.472	1.654
2	23.157	9.241	3.770	2.157	1.920	2.005	2.104
3	60.425	21.415	8.066	4.425	3.686	3.626	3.677
4	115.83	38.259	13.882	7.482	6.080	5.863	5.877
5	188.95	59.654	21.220	11.337	9.099	8.687	8.659
10	809.31	233.54	80.707	42.600	33.508	31.472	31.083
20	3273.9	912.47	313.68	165.12	128.84	120.20	118.26
30	7341.7	2031.7	698.66	367.64	286.18	266.44	261.82
100	80436.	22188.	7649.5	4025.3	3123.6	2900.1	2845.3

Thus

$$\mu_2 = \frac{1}{2T} > 0$$

(bifurcation occurs for $r > \frac{2}{T}$),

$$\beta_2 = \frac{-1}{10T} < 0$$

(the bifurcating periodic solutions are asymptotically, orbitally stable), and

$$\tau_2 = 1/12 > 0 .$$

Furthermore, the asymptotic form of the bifurcating periodic solution is

$$x(t,r) = 2[2T(r - \tfrac{2}{T})]^{1/2} \cos \tfrac{t}{T} + 2T(r - \tfrac{2}{T})[\tfrac{7}{6} \cos (\tfrac{2t}{T}) + \tfrac{1}{6} \sin(\tfrac{2t}{T})]$$

$$+ O([r - \tfrac{2}{T}]^{3/2}) .$$

7. EXERCISES

<u>Exercise 1</u>. Show that for (3.18) - (3.19) the Hopf spectral

hypotheses are met for $\tau = \tau_c$, where

$$\tau_c = \frac{\text{Arc cos } (\frac{-a}{b})}{(b^2 - a^2)^{1/2}} \qquad (0 < \text{Arc cos } (\frac{-a}{b}) < \pi) .$$

The steady state x_e is stable for $\tau < \tau_c$ and loses stability as τ increases through τ_c .

Exercise 2. Investigate the stability of the periodic solutions of

$$\dot{u}(t) = -au(t) - bu(t - \tau) + cu^2(t - \tau) + du^3(t - \tau)$$

$$(u \text{ a scalar})$$

that arise from Hopf bifurcation with τ as the bifurcation parameter, and show that if $d > 0$, then $\beta_2 < 0$ (at least in the special case $a = 0$).

Exercise 3. Compute $c_1(0)$ for (3.18) - (3.19) (at least if $a = 0$). Hint: Expand the function Λ in (3.19) about the equilibrium x_e, and note that terms of order 4 and higher do not affect the bifurcation calculations (if only terms through $O(\epsilon^2)$ are desired).

Exercise 4. Perform the reduction of (6.4) to a system of ordinary differential equations and by analyzing the system confirm the results of Section 6. Hint: Show that if one defines

$$Y_1(t) = \int_{-\infty}^{0} N(t+s)Q(-s)ds$$

$$\text{and} \quad Y_2(t) = \dot{Y}_1(t) ,$$

then the system (6.1) becomes

$$\dot{N} = rN(1 - Y_1) - \alpha PN ,$$

$$\dot{P} = -bP + \beta PN ,$$

$$\dot{Y}_1 = Y_2 ,$$

$$\dot{Y}_2 = N/T^2 - 2Y_2/T - Y_1/T^2$$

since

$$\ddot{Q} + 2\dot{Q}/T + Q/T^2 = 0 .$$

This reduction of (6.1) to a system of ordinary differential equations illustrates a popular technique; see, for example, MacDonald [77]. (We have verified Table 6.1 by applying BIFOR2 to the above 4×4 system; see subroutine MA on the microfiche.)

Indeed, if a system of integrodifferential equations is given in which each integral that occurs has the form

$$(*) \qquad \int_{-\infty}^{0} u(t+s)Q(-s)ds ,$$

and Q satisfies a constant coefficient scalar differential equation of order n, then by defining n new variables (for each distinct integral (*)) Y_i $(i = 1,\ldots,n)$ to be

$$Y_i(t) = \int_{-\infty}^{0} u(t+s)Q^{(i-1)}(-s)ds ,$$

one is able to reduce the system of integrodifferential equations to a much larger system of ordinary differential equations.

<u>Exercise 5</u>. This exercise illustrates a technique by which delay differential equations are approximated by systems of ordinary differential equations.

Consider the Hutchinson-Wright equation

$$\dot{x} = -a\,x(t-1)\,(1+x(t)) . \qquad \text{(H-W)}$$

Define

$$Q_n(t) = n^{n+1} t^n e^{-nt}/n! \,, \qquad (n = 1,2,\ldots) \,.$$

Suppose that $x(t)$ is bounded and C^1 on $-\infty < t < \infty$. Convince yourself that for any fixed t,

$$\lim_{n\to\infty} \int_{-\infty}^{t} Q_n(t-s)x(s)ds = x(t-1) \,.$$

Thus (H-W) may be approximated by the integro-differential equation

$$x = -a\left[\int_{-\infty}^{t} Q_n(t-s)x(s)ds\right](1+x(t)) \,. \qquad (HW_n)$$

Show that (HW_n) is equivalent to the $n+2$ dimensional system of ordinary differential equations

$$\dot{y}_1 = -ay_2(1+y_1)$$

$$\dot{y}_i = y_{i+1} \qquad (2 \le i \le n+1)$$

$$\dot{y}_{n+2} = \Sigma_{j=1}^{n+2} q_j y_j \,,$$

where

$$q_1 = n^{n+1} \,,$$

$$q_j = -C_{n+1,j-2}\, n^{n+3-j} \qquad (2 \le j \le n+2) \,.$$

Here $C_{n,k} = n!/((n-k)!k!)$.

Hint. $Q_n(t)$ satisfies $(d/dt + n)^{n+1} Q_n = 0$.

Remark. For $n = 1,2,\ldots$ each system HW_n exhibits a Hopf bifurcation as the parameter a is increased past a critical value a_c^n. The sequence $\{a_c^n\}$ converges from above to $\pi/2$. We have used this technique (see subroutine HW on microfiche) to verify the hand calculations for (H-W). Although the convergence of the method is slow. Richardson extrapolation, based upon an assumed functional dependence

$$a_c^n = a_c + \Sigma_{j=1}^4 c_j/n^j + O(n^5) ,$$

yields good agreement with $\pi/2$. A similar analysis holds for α_2, β_2, and τ_2 . We have used this same technique to verify the hand calculations for the delay equation with two distinct delays studied in Section 4 (see subroutine KY on microfiche).

Exercise 6. Show that in each of the examples considered in Sections 3 and 4 the eigenvalues of the linear problem with negative real parts are uniformly bounded away from the imaginary axis.

CHAPTER 5. APPLICATIONS: PARTIAL DIFFERENTIAL EQUATIONS (BY HAND)

1. INTRODUCTION

In many chemical, physical and biological problems, the state of the system can be described by functions and the dynamics by partial differential equations of evolution. Often the system involves certain external parameters which can be varied or controlled. We investigate the situation in which a single parameter ν is being changed.

Suppose the system has a stationary state for parameters in an interval, and suppose the linear stability of the stationary state changes at a certain critical value ν_c in the interval. Provided the appropriate Hopf conditions are satisfied, a periodic solution branches from the stationary state at ν_c, i.e. there is a Hopf bifurcation for the partial differential system. A Hopf bifurcation theorem can be proved when the systems are of a certain "parabolic" type. This has been done by Iooss [59], Judovich [67], Joseph-Sattinger [66], Crandall and Rabinowitz [26], and others. Here, we shall again apply the method which we used in the previous chapters in dealing with ordinary differential equations and delay equations: we shall apply the center manifold theorem to reduce the bifurcation problem to one for ordinary differential equations on the center manifold. This reduced bifurcation problem was solved in detail in Chapter 1. The idea of using the center manifold theorem to study bifurcation problems can be found in Ruelle-Takens [97] and Marsden [80]. The existence of center manifolds for partial differential equations of evolution is proved in Marsden-McCracken [81] or Iooss [61]. It is observed that partial differential equations of evolution can define a local smooth semiflow (in nice cases). Let the local smooth semiflow be expressed as a collection of maps F_t. One still has center manifolds for maps, which are not necessarily diffeomorphisms. The desired center manifold is a center manifold of F_1. For more details about these results, one may consult Appendix A, as well as [53, 61, 81].

Because of the extreme generality of center manifold theory, individual applications tend to be involved. One has the usual difficulty of verifying that the general hypotheses of the theory are fulfilled. In the future we expect to see various bifurcation theorems, based upon center manifold theory, but specialized for easy application.

This final chapter is arranged as follows: In Section 2 local semiflows are defined, and we show how they can arise from certain reaction-diffusion systems. This discussion includes, in particular, applications of the theory of linear semigroups and a result originally due to Segal [99]. A theory of Hopf bifurcation for local semiflows and algebraic formulae for stability computations are given in Section 3. This theory applies immediately to reaction diffusion systems on smooth, bounded domains in $\mathbb{R}^n$ $(n \le 3)$ with Neumann or Dirichlet boundary conditions. Two examples, illustrating the theory and computations of bifurcation formulae, are presented in Sections 4 and 5. For the reader's convenience, some results from semigroup theory and function theory in Sobolev space are collected in Appendices B and C.

Before proceeding, we acknowledge our debtedness to Jerry Marsden. This Chapter is an exposition - application of his center manifold theory for semiflows.

2. SEMIFLOWS

Set $\mathbb{R}_+ = \{t \in \mathbb{R} | t \geq 0\}$. A C^o (i.e. continuous) local semiflow on a Banach space X is a continuous map $F: \mathcal{D} \subset \mathbb{R}_+ \times X \to X$, where $\mathcal{D}$ is an open set in $\mathbb{R}_+ \times X$ with $\{0\} \times X \subset \mathcal{D}$, such that $F(0,x) = x$ for all x in X and $F(s+t, x) = F(s, F(t,x))$ for any $s,t \geq 0$, $x \in X$, whenever both sides of this equations are defined.

For each $t \geq 0$, let $\mathcal{D}_t$ be the open set $\{x \in X | (t,x) \in \mathcal{D}\}$ in X , and define $F_t: \mathcal{D}_t \to X$ by $F_t(x) = F(t,x)$ for all x in $\mathcal{D}_t$. The local semiflow is called a C^o linear semiflow (or a C^o linear semigroup) if each F_t is a bounded linear transformation on X ; a C^r local semiflow, if each F_t is of class C^r . Clearly, any C^o linear semiflow is a local C^r semiflow (in x) for any $r \geq 0$. (The concept of a C^r local semiflow extends naturally to Banach manifolds.)

We shall first demonstrate by means of examples how the theory of linear semigroups (Appendix B) may be used to establish that linear partial differential equations of evolution, $du/dt = Lu$, can naturally give rise to linear semiflows. This amounts to proving that the linear operator L (in the appropriate function space) is indeed the infinitesimal generator for some semigroup.

Denote by $x = (x_1,\ldots,x_n)$ a variable point in $\mathbb{R}^n$, and let

$$D^\alpha = \frac{\partial^{\alpha_1}}{\partial x_1^{\alpha_1}} \cdots \frac{\partial^{\alpha_n}}{\partial x_n^{\alpha_n}} ,$$

where $\alpha = (\alpha_1,\ldots,\alpha_n)$, $|\alpha| = \alpha_1 + \ldots + \alpha_n$. Given an open bounded domain Ω in $\mathbb{R}^n$ with smooth boundary $\partial\Omega$, $C^m(\Omega)$ $(C^m(\bar{\Omega}))$ stands for the set of complex-valued smooth functions of class C^m on $\Omega(\bar{\Omega})$. For $u \in C^m(\Omega)$, set $|u|_m^\Omega = [\int_\Omega \sum_{|\alpha| \leq m} |D^\alpha u|^2 dx]^{\frac{1}{2}}$. The Sobolev space $H^m(\Omega)$ is the completion of the space $\{u \in C^m(\Omega) \mid |u|_m^\Omega < \infty\}$ under the norm $|\ |_m^\Omega$. The famous compactness theorem of Rellich [95], states that the imbedding $H^m(\Omega) \to H^j(\Omega)$ with $j < m$ is compact. Set $H^\ell(\Omega, \mathbb{R}) = \{u \in H^\ell(\Omega) \mid u \text{ is real}\}$. Let $f\colon \Omega \times \mathbb{R}^r \to \mathbb{R}$ be smooth with $\ell > \frac{n}{2}$. It is shown in Appendix C that the map $(s_1,\ldots,s_r) \to f(\cdot, s_1(\cdot),\ldots,s_r(\cdot))$ from $\oplus^r H^\ell(\Omega,\mathbb{R}) \to H^\ell(\Omega,\mathbb{R})$ is well-defined and smooth.

Examples

(a) Consider the Dirichlet problem:

$$\Delta u = \left(\frac{\partial^2}{\partial x_1^2} + \ldots + \frac{\partial^2}{\partial x_m^2}\right) u = f$$

in Ω with $u = 0$ on $\partial\Omega$. For $u \in C^{k+2}(\bar{\Omega})$ with $u = 0$ on $\partial\Omega$, one can establish the a priori estimate:

$$|u|_{2+k}^\Omega \le C(|\Delta u|_k^\Omega + |u|_0^\Omega) \quad (\text{in fact, } |u|_{2+k}^\Omega \le C|\Delta u|_k^\Omega) .$$

This a priori inequality is based on Gårding's inequality for the Laplacian (see Friedman [34]). By standard theory [85, Chapters 7 and 8], the Laplacian can be extended to a closed, self-adjoint operator $A\colon D_A \to L^2(\Omega) \equiv H^0(\Omega)$ with dense domain D_A consisting of the closure in $H^2(\Omega)$ of the set

$$C_0^2(\bar{\Omega}) = \{u \in C^2(\bar{\Omega}) \mid u = 0 \text{ on } \partial\Omega\} .$$

The space $L^2(\Omega)$ is a Hilbert space and A is dissipative since for $u \in C_0^2(\bar{\Omega})$, $\langle \Delta u,u\rangle \le 0$, which implies $\langle Au,u\rangle \le 0$ for $u \in D_A$. Therefore, by Theorem 3 in Appendix B, A generates a contraction semigroup.

It can be proved that for any ϕ $(0 < \phi < \pi/2)$ there exists an $a > 0$ such that the resolvent $(\lambda I - A)^{-1}$ exists for all λ in $S_{\phi,a} = \{\lambda \in \mathbb{C} \mid \lambda \ne a \text{ and } |\arg(\lambda - a)| < \phi + \pi/2\}$. Moreover $\|(\lambda I - A)^{-1}\| \le C/|\lambda|$ for some $C > 0$ and $\lambda \in S_{\phi,a}$. These results also hold for more general second order elliptic operators [85]. Thus, it follows by Theorem 4 in Appendix B that A generates an analytic semigroup T_t. In particular, $T_t u \in D_A$ for any $t > 0$, $u \in L^2(\Omega)$, and $\|T_t\|$, $\|AT_t\|$ are continuous in t for $t > 0$. Given any λ in the resolvent set $\rho(A)$ of A, $(\lambda I - A)^{-1}(L^2(\Omega)) \subset H^2(\Omega)$. By Rellich's compactness theorem, $(\lambda I - A)^{-1}$ is compact. Hence, by Theorem 5 in Appendix B, one concludes that A generates a compact semigroup.

The domain D_A of A becomes a Banach space under the norm $|\ |_2^\Omega$. For future use we let $\tilde{A}$ be the restriction of A to the subspace $D_{A^2} = \{u \in D_A | Au \in D_A\}$. Since $T_t u \in D_A$ if $u \in D_A$, we may also let $\tilde{T}_t$ be the restriction of T_t to the subspace D_A of $L^2(\Omega)$, and we obtain a C^o semigroup $\tilde{T}_t$ on D_A (recall that $|u|_2^\Omega \le C(|\Delta u|_0^\Omega + |u|_0^\Omega)$ for $u \in D_A$). The C^o semigroup $\tilde{T}_t$ is compact and has infinitesimal generator $\tilde{A}$.

(b) Consider the Neumann problem: $\Delta u = f$ with $\partial u/\partial n = 0$ on $\partial\Omega$. Here $\partial/\partial n$ denotes the directional derivative of u in the outward normal direction on $\partial\Omega$. For $u \in C^2(\bar{\Omega})$ with $\partial u/\partial n = 0$ on $\partial\Omega$, one can establish the a priori estimate: $|u|_2^\Omega \le C(|\Delta u|_0^\Omega + |u|_0^\Omega)$ as in (a).
Denote by D_B, the dense subset of $L^2(\Omega)$, which consists of functions belonging to the closure of the set $\{u \in C^2(\bar{\Omega}) | \ \partial u/\partial n = 0 \text{ on } \partial\Omega\}$ in $H^2(\Omega)$. Again, the Laplacian can be extended to a closed, selfadjoint operator $B: D_B \to L^2(\Omega)$ [85, pp. 237-240] that generates an analytic contraction semigroup T_t on $L^2(\Omega)$, where each T_t is compact for $t > 0$. The space D_B becomes a Banach space under the norm $|\ |_2^\Omega$. As above, we restrict B on this space D_B and call it $\tilde{B}$; namely, $\tilde{B}: D_{B^2} \to D_B$. Let $\tilde{T}_t$ be the restriction of T_t on D_B. It is not hard to show that $\tilde{T}_t$ is again a C^o compact semigroup with infinitesimal generator $\tilde{B}$.

Next, we want to show how local semiflows (nonlinear) arise from nonlinear partial differential equation of evolution in a special situation. We follow the presentation in Holmes and Marsden [53]. Let L be the generator of a C^o semigroup U_t on X, and $f: X \to X$ be a map of class $C^k (k \geq 1)$. Now, consider the abstract evolution system $du/dt = Lu + f(u)$. Clearly, any solution $u(t)$ for this system with $u(0) = u_o \in D_L$ satisfies the integral equation (Duhamel's formula)

$$u(t) = U_t u_o + \int_0^t U_{t-s} f(u(s))ds \qquad (\dagger)$$

Since f is locally Lipschitz and $\|U_t\| \leq Me^{t\beta}$ for some constants M, β. Picard iteration, as for ordinary differential equations, shows that the solutions of the integral equation (†) define a unique local semiflow $F: \mathcal{D} \subset \mathbb{R}_+ \times X \to X$. $\mathcal{D}$ is defined as follows. For each u_o in X, let $J(u_o)$ be the maximal interval of the form $[0,\alpha)$, where the solution of the integral equation with $u(0) = u_o$ exists. This solution on $J(u_o)$ with initial condition $u(0) = u_o$ is unique. $\mathcal{D}$ is given as $\bigcup_{u_o \in X} (J(u_o), u_o)$ and $t \to F(t,u_o)$ is the unique solution on $J(u_o)$. Furthermore, F_t $(F_t(u) = F(t,u))$ is of class C^k in u for each $t > 0$. Let us call F the local semiflow associated to the abstract differential equation $du/dt = Lu + f(u)$. If $f(0) = 0$ and $Df(0) = 0$, then $U_t = D_x F_t(0)$ for all $t \geq 0$.

In the bifurcation problem setting, L and f in the equation $du/dt = Lu + f(u)$ depend on some parameter ν lying in an open interval containing ν_c. Define $\mu = \nu - \nu_c$. Thus, one has $du/dt = L_\mu u + f_\mu(u)$. Denote by U_t^μ, F_t^μ the corresponding semigroup and local semiflow of L_μ and $L_\mu + f_\mu$ respectively. Now, assume

(1) U_t^μ is jointly continuous in t, u, μ $(t \geq 0)$,

(2) U_t^μ is of class C^k in u,μ for each $t > 0$.

(3) f_μ is of class C^k in u,μ .

Under the hypotheses (1), (2) and (3) on L_μ and f_μ , it follows by the same arguments as above that F_t^μ is of class C^k in u,μ for each $t > 0$.

(c) Consider a reaction-diffusion system:

$$\frac{du_i}{dt} = d_i \Delta u_i + \Sigma_{j=1}^m c_{ij} u_j + f_i(u) \qquad (i = 1,\dots,m)$$

on a bounded domain Ω in $\mathbb{R}^n$ $(n \leq 3)$ with Dirichlet boundary condition $u = 0$ on the smooth boundary $\partial\Omega$. Here, d_i, c_{ij} are real numbers $d_i > 0$, $f_i : \mathbb{C}^m \to \mathbb{C}$ are smooth functions in $u_1,\dots,u_m$ with $f_i(0) = 0$, $D_i f(0) = 0$ and $f_i(u)$ is real for real $u_1,\dots,u_m$. We write this system in the more compact form: $du/dt = D\,\Delta u + Cu + f(u)$ in Ω with $u = 0$ on $\partial\Omega$. Here

$$u = \begin{pmatrix} u_1 \\ \vdots \\ u_m \end{pmatrix}, \quad D = \begin{pmatrix} d_1 & \dots & 0 \\ \vdots & & \vdots \\ 0 & \dots & d_m \end{pmatrix}, \quad C = (c_{ij}) \text{ , and } \quad f = \begin{pmatrix} f_1 \\ \vdots \\ f_m \end{pmatrix}$$

Denote by

$$\tilde{A}_i : D_{A_i^2} \to D_{A_i}$$

the operator which is associated with the Dirichlet problem for the operator $d_i \Delta$ as in a). Recall that D_{A_i} is the closure of the set $\{u_i \in C^2(\bar{\Omega}) | u_i = 0$ on $\partial\Omega\}$ in $H^2(\Omega)$. Set $X = D_{A_1} \times \ldots \times D_{A_m}$ and $\tilde{A} = \tilde{A}_1 \times \ldots \times \tilde{A}_m$. Clearly $\tilde{A}$ generates a compact analytic semigroup on X. Let $L = \tilde{A} + C$ be a perturbation of $\tilde{A}$ by a bounded operator C. By Theorem 6 stated in Appendix B, L again generates a compact, analytic semigroup. From the remark made just before Example (a) (here $n \le 3$), it follows that $f(u)$ defines a smooth function from X to X. Therefore, the Dirichlet problem for the reaction-diffusion system above defines a smooth local semiflow on X.

Now, we assume $D = D_\mu$ and $C = C_\mu$ depend analytically on a real parameter μ near 0. Clearly, the family $L_\mu = D_\mu \Delta + C_\mu$ of closed linear operators on X can be extended naturally into a holomorphic family of closed linear operators of type (A) near 0. Thus, by Theorem 7 in Appendix B, the semiflow U_t^μ is jointly continuous in t, u, μ and U_t^μ is smooth in u, μ for each $t > 0$. Let us assume that $f = f(u, \mu)$ also depends on the parameter μ and $f(u, \mu)$ is smooth in u, μ. Consequently, the equation $du/dt = D_\mu \Delta u + C_\mu u + f(u, \mu)$ with $u = 0$ on $\partial\Omega$ defines a local smooth semiflow F_t^μ on X enjoying the same smoothness of that of U_t^μ.

In many situations, u only takes real values. Set $Y = \{u \in X | u$ is real$\}$, the real subspace of X. The local semiflow F_t^μ, defined through Picard iterations, leaves Y invariant; each Picard iterate maps Y to Y. In other words, F_t^μ can be regarded as a local semiflow on Y, and it carries the same smoothness as that of F_t^μ on X.

(d) Similarly, the reaction-diffusion equation $du/dt = D \Delta u + Cu + f(u)$ in Ω with Neumann condition $\partial u/\partial n = 0$ on $\partial\Omega$, also generates a local smooth semiflow on X or Y. Here, u, D, C, f have the same meaning as that in (c),

$X = D_{B_1} \times \dots \times D_{B_m}$, $\tilde{B}_i$ is the infinitesimal generator in $D_{\tilde{B}_i}$ introduced for the Neumann problem in (b), and $Y = \{u \in X | u \text{ is real}\}$. If D, C, f depend on a real parameter μ near 0 in the same way as in (c), then the associated local semiflow F_t^μ on X or Y has the same smoothness as that in (c). Now $\tilde{B} = \tilde{B}_1 \times \cdots \times \tilde{B}_m$ and $L = \tilde{B} + C$.

(e) Suppose the bifurcation of interest is from a non-trivial stationary solution u_* of the system

$$\frac{du}{dt} = D\Delta u + Cu + f(u) \qquad (x \in \Omega \subset \mathbb{R}^n)$$

with either Dirichlet or Neumann conditions as in (c) or (d). Set $v = u - u_*$. Then

$$\frac{dv}{dt} = D\Delta v + C_*(x)v + g(x,v) ,$$

where

$$C_*(x) = C + f_x(u_*) ,$$

$$g(x,v) = f(u_* + v) - f(u_*) - f_u(u_*)v .$$

Provided u_* belongs to the appropriate Banach space and is smooth in μ , the analysis of (c) or (d) carries through with minor modifications. Consequently, the system for v with either Dirichlet or Neumann conditions on $\partial\Omega$ defines a local smooth semiflow F_t^μ with the same smoothness properties as in (c) or (d).

For further examples of nonlinear semiflows see Holmes-Marsden [53] and Marsden-McCracken [81]. Indeed bifurcation in reaction-diffusion systems has been studied by a large number of authors; see Paul Fife's recent surveys [31; 32, pp. 152-154]. The <u>Proceedings</u> of the October 1979 Madison, Wisconsin Advanced Symposium on Dynamics and Modelling of Reactive Systems [102] deal with reaction-diffusion in general.

3. HOPF BIFURCATION AND ASSOCIATED STABILITY COMPUTATIONS FOR LOCAL SEMIFLOWS

Let Y be a Banach space admitting a C^∞ norm away from 0. Now, consider a family of C^0 local semiflows F_t^μ defined in a neighborhood of 0 in Y for $0 \le t \le \tau$ and μ near 0 in $\mathbb{R}$ with $F_t^\mu(0) = 0$. These semiflows F_t^μ may come from some partial differential equation of evolution. The suspended local semiflow $Y \times \mathbb{R} \to Y \times \mathbb{R}$ of F_t^μ is defined by $(y,\mu) \to (F_t^\mu(y),\mu)$. To study Hopf bifurcation for F_t^μ we need some hypotheses:

Smoothness hypothesis. $F_t^\mu(y)$ is jointly continuous in t,y,μ $(t \ge 0)$ and to each $t > 0$, $F_t^\mu(y)$ is of class C^{k+1} in y,μ $(k \ge 5)$.

Spectral hypothesis. (1) $F_t^\mu(0) = 0$ for all t in $[0,\tau]$ and μ near 0. (2) The semigroup $D_Y F_t^\mu(0)$ has infinitesimal generator A_μ and $\exp(t\sigma(A_\mu)) = \sigma(D_Y F_t^\mu(0))\backslash\{0\}$ for all $t > 0$. (3) A_μ has a pair of simple complex conjugate eigenvalues $\lambda(\mu), \overline{\lambda(\mu)}$ with $\lambda(0) = i\omega$ $(\omega_0 > 0)$ and $d\mathrm{Re}\lambda(\mu)/d\mu|_{\mu=0} > 0$. (4) There exists $\delta > 0$ such that $\mathrm{Re}[\sigma(A_\mu)\backslash\{\lambda(\mu), \overline{\lambda(\mu)}\}] \le -\delta < 0$ for μ near 0.

We assume F_t^μ satisfies the smoothness and the spectral hypotheses as described above. By the center manifold theorem for semiflows (see Appendix A), there exists a locally invariant, locally attractive, three-dimensional submanifold $\mathfrak{m}$ of the suspended local semiflow $(y,\mu) \to (F_t^\mu(y),\mu)$ in $Y \times \mathbb{R}$ through $(0,0)$. This manifold (called the center manifold) is tangent to $Y_c \times \mathbb{R}$, where Y_c is the eigenspace associated with the eigenvalues $i\omega_0$, $-i\omega_0$. The existence of center manifolds for the partial differential equations studied in this Chapter (i.e. reaction-diffusion equations) also follows readily from

Appendix A. For all sufficiently small μ, let M_μ denote the slice μ = constant of the center manifold $\mathfrak{m}$. Then $\mathfrak{m}$ is locally of the form

$$\mathfrak{m} = \{(M_\mu(y),\mu) \mid \|y\| < \epsilon,\ |\mu| < \delta,\ (y,\mu) \in Y \times \mathbb{R}\}$$

for some small $\epsilon > 0$ and $\delta > 0$.

Next we need a result of Chernoff and Marsden [12; 17; 81, p. 265] which is a generalization of the Bochner-Montgomery theorem:

<u>Theorem</u>. <u>Let</u> G <u>be a local</u> C^k <u>semiflow on a finite dimensional smooth manifold</u> $\mathfrak{m}$. <u>Then</u>, G <u>is locally reversible</u>, <u>is jointly</u> C^k <u>in</u> t,y, <u>and is generated by a</u> C^{k-1} <u>vector field on</u> $\mathfrak{m}$.

By this result, the restriction to $\mathfrak{m}$ of the suspended local semiflow $G: (y,\mu) \to (F_t^\mu(y),\mu)$ is generated by a vector field R. Take any complementary space Y_s of Y_c in Y, and write $Y = Y_c \oplus Y_s$. Clearly, the projection $(y_c,y_s,\mu) \to (y_c,\mu)$ defines a coordinate system on $\mathfrak{m}$. In this coordinate system, the vector field R is actually a family of vector fields R_μ on Y_c parameterized by μ. Since all local recurrence of G happens in the center manifold $\mathfrak{m}$, the bifurcation problem for F_t^μ reduces to that for R_μ. The vector fields R_μ can be regarded as the "essential model" for the semiflow G near $(0,0)$.

The smoothness and spectral hypotheses on F_t^μ imply that R_μ obeys the same kind of hypotheses. By Theorem 2 in Chapter 1, we have a Hopf Bifurcation of R_μ or F_t^μ at $\mu = 0$. Therefore, F_t^μ has a family of periodic solutions $p(t,\mu)$ with period T_μ, which can be parametrized by ϵ near 0:

$$\mu = \mu(\varepsilon) = \mu_2\varepsilon^2 + O(\varepsilon^4)$$

$$T = T(\varepsilon) = \frac{2\pi}{\omega_0}\ [1 + \tau_2\varepsilon^2 + O(\varepsilon^4)]\ .$$

ny periodic solutions sufficiently close to the origin in Y ith $|\mu|$ small must appear in this family.

It will be shown below that the differential equation escribing R_0 in Y_c can be written in the form:

$$\frac{dz}{dt} = i\omega_0 z + \sum_{2\le i+j\le 3} \frac{g_{ij}}{i!j!} z^i \bar{z}^j + O(|z|^4) \quad \text{with} \quad g_{ij} \in \mathbb{C}\ ,$$

hen one identifies Y_c with $\mathbb{C}$ through a wise choice of oordinates.

For the convenience of the reader, the bifurcation formulae or μ_2, τ_2 in terms of g_{ij} and $\lambda'(0)$, which have been btained in Chapter 1, Section 3, are collected here:

$$\mu_2 = -\frac{\operatorname{Re} c_1(0)}{\alpha'(0)}$$

$$\tau_2 = -\frac{1}{\omega_0}\ [\operatorname{Im} c_1(0) + \mu_2\omega'(0)]\ ,$$

where $\lambda(\mu) = \alpha(\mu) + i\omega(\mu)$

and $c_1(0) = \frac{i}{2\omega_0}(g_{20}g_{11} - 2|g_{11}|^2 - \frac{1}{3}|g_{02}|^2) + \frac{g_{21}}{2}\ .$

table periodic solutions bifurcate if $\operatorname{Re} c_1(0) < 0$ and istable periodic solutions bifurcate if $\operatorname{Re} c_1(0) > 0$. Here, ie must use the local attractivity of the center manifold and ote that for partial differential equations this implies global cistence of orbits near any stable periodic solution in a enter manifold; see Appendix A.

For the rest of this section, we show how the differential equation describing R_0 on the manifold M_0 can be obtained in explicit form for a local semiflow defined by a partial differential system of a specific type. The process will be formally the same as that used in previous Chapters.

At $\mu = 0$, our partial differential system has the abstract form $du/dt = Au + f(u)$ on a Banach space $Y \oplus iY = \{y_1 + iy_2 | y_1, y_2 \in Y\}$, where A generates a C^0 semi-group and f is of class C^{k+1} with $f(0) = 0, Df(0) = 0$ $(k \geq 5)$, and $Au, f(u) \in Y$ for $u \in Y$. Since Hopf bifurcations are being considered, we know that A has a pair of simple eigenvalues $\pm i\omega_0$, $\omega_0 > 0$, with eigenvectors $q, \bar{q}$.

Suppose now: (1) There is a real inner product $\langle\ ,\ \rangle$ on Y, which is continuous with respect to the topology on Y, and it is extended to $Y \oplus iY$ so that $\langle \lambda u, v\rangle = \bar{\lambda}\langle u, v\rangle$; (2) There exists a closed operator A^* in $Y \oplus iY$ such that $\langle u, Av\rangle = \langle A^* u, v\rangle$; and (3) There exists q^* in $Y \oplus iY$ such that $A^* q^* = -i\omega_0 q^*$, $\langle q^*, q\rangle = 1$ and $\langle q^*, \bar{q}\rangle = 0$.

Write $Y = Y_c \oplus Y_s$, where $Y_c = \{zq + \bar{z}\bar{q} | z \in \mathbb{C}\} \approx \mathbb{C}$ and $Y_s = \{u \in Y | \langle q^*, u\rangle = 0\}$. Thus, any element u in Y can be identified with a pair (z, w) such that $u = zq + \bar{z}\bar{q} + w$ with $z \in \mathbb{C}$ and $w \in Y_s$. The differential equation in (z, w) becomes:

$$\frac{dz}{dt} = i\omega_0 z + \langle q^*, f(zq + \bar{z}\bar{q} + w)\rangle \tag{5.1}$$

$$\frac{dw}{dt} = Aw + H(z, \bar{z}, w), \tag{5.2}$$

where $H(z, \bar{z}, w) = f - \langle q^*, f\rangle q - \langle \bar{q}^*, f\rangle \bar{q}$.

Write

$$H(z, \bar{z}, w) = \frac{H_{20}}{2} z^2 + H_{11} z\bar{z} + \frac{H_{02}}{2} \bar{z}^2 + O(|z|^3) + O(|z||w|) \tag{5.3}$$

All of H_{20}, H_{11}, H_{02} belong to Y_s. The manifold M_0 can be expressed as the graph of a C^k function $w: Y_c \approx C \to Y_s$. Set

$$w = \frac{w_{20}}{2} z^2 + w_{11} z \bar{z} + \frac{w_{02}}{2} \bar{z}^2 + O(|z|^3) . \tag{5.4}$$

Since the local semiflow restricted to the manifold M_0 is generated by a C^{k-1} vector field on M_0, $t \to F_t(u)$ satisfies the equation $du/dt = Au + f(u)$ in Y at $\mu = 0$ for u in M_0. Hence,

$$Aw + H(z,\bar{z},w) = \frac{dw}{dt} = \frac{\partial w}{\partial z}\frac{dz}{dt} + \frac{\partial w}{\partial \bar{z}}\frac{d\bar{z}}{dt} .$$

Now substitute H from equation (5.3), dz/dt from equation (5.1) and w from equation (5.4) into the equation

$$Aw + H = \frac{\partial w}{\partial z}\frac{dz}{dt} + \frac{\partial w}{\partial \bar{z}}\frac{d\bar{z}}{dt} ,$$

and then expand in powers of $z, \bar{z}$. One gets

$$A\{\frac{w_{20}}{2} z^2 + w_{11} z \bar{z} + \frac{w_{02}}{2} \bar{z}^2 + \ldots\} + \{\frac{H_{20}}{2} z^2 + H_{11} z \bar{z} + \frac{H_{02}}{2} \bar{z}^2 + \ldots\}$$

$$= (w_{20} z + w_{11}\bar{z} + \ldots)(i\omega_0 z + \ldots) + (w_{11} z + w_{02}\bar{z} + \ldots)(-i\omega_0 \bar{z} + \ldots).$$

The z^2, $z\bar{z}$, $\bar{z}^2$ terms of the above equation give

$$(2i\omega_0 - A)w_{20} = H_{20} ,$$

$$- Aw_{11} = H_{11} ,$$

$$(-2i\omega_0 - A)w_{02} = H_{02} .$$

Thus

$$w_{20} = (2i\omega_0 - A)^{-1} H_{20} ,$$

$$w_{11} = -A^{-1} H_{11} ,$$

$$w_{02} = \overline{w}_{20} .$$

Since 0 and $2i\omega_0$ do not belong to the spectrum of A, by our hypotheses, the inverses $(2i\omega_0 - A)^{-1}$, A^{-1} do exist.

To obtain the equation on M_0 in terms of the $z, \overline{z}$ coordinates up to the third order, it suffices to replace w in the equation (5.1) by

$$(2i\omega_0 - A)^{-1} H_{20} \frac{z^2}{2} + (-A^{-1} H_{11}) z\overline{z} + (-2i\omega_0 - A)^{-1} H_{02} \frac{\overline{z}^2}{2} + \dots .$$

Thus the differential equation on M_0 appears in the form:

$$\frac{dz}{dt} = i\omega_0 z + \sum_{2 \le i+j \le 3} \frac{g_{ij}}{i!j!} z^i \overline{z}^j + O(|z|^4) .$$

4. THE BRUSSELATOR WITH DIFFUSION AND NO FLUX BOUNDARY CONDITION ON AN INTERVAL OR A DISK

In this section, we first study the Hopf bifurcation in the Brusselator (the Lefever-Prigogine reaction-diffusion system in [7, 11, 75]), on the unit interval with Neumann (no flux) boundary conditions. The system is

$$\frac{\partial X}{\partial t} = A - (B + 1)X + X^2 Y + d\,\frac{\partial^2 X}{\partial r^2}$$

$$\frac{\partial Y}{\partial t} = BX - X^2 Y + \theta\, d\,\frac{\partial^2 Y}{\partial r^2}$$

with boundary conditions

$$\frac{\partial X}{\partial r}(0,t) = \frac{\partial X}{\partial r}(1,t) = 0$$

$$\frac{\partial Y}{\partial r}(0,t) = \frac{\partial Y}{\partial r}(1,t) = 0 \ .$$

Here r belongs to $[0,1]$, X,Y are real-valued functions depending on two variables t and r, A,d,θ are positive constants, and $B > 0$ is taken as the bifurcation parameter ν. The significance of this system is explained in Chapter 2, Example 3.

A stationary solution is given by $(A, B/A)$. To shift this stationary solution to the origin, we let

$$u = X - A$$

$$v = Y - B/A \ .$$

Then,

$$\frac{\partial u}{\partial t} = (B - 1)u + A^2 v + d \frac{\partial^2 u}{\partial r^2} + h(u,v)$$

$$\frac{\partial v}{\partial t} = -Bu - A^2 v + \theta d \frac{\partial^2 v}{\partial r^2} - h(u,v) ,$$

with

$$\frac{\partial u}{\partial r}(0,t) = \frac{\partial u}{\partial r}(1,t) = 0$$

$$\frac{\partial v}{\partial r}(0,t) = \frac{\partial v}{\partial r}(1,t) = 0 .$$

Here, $h(u,v) = \frac{B}{A} u^2 + 2Auv + u^2 v$.

By example (d) in Section 2 of this Chapter, the Brusselator generates a smooth local semiflow F_t on $Y = \{(u,v) \in D_B \times D_B \mid u,v$ are real$\}$, where $D_B = \{u \in H^2[0,1] \mid \frac{\partial u}{\partial r} = 0$ at $0,1\}$. The object $D_X F_t(0)$ is a compact semigroup with infinitesimal generator L given by

$$L = \begin{pmatrix} (B - 1) + d \frac{\partial^2}{\partial r^2} & A^2 \\ - B & - A^2 + \theta d \frac{\partial^2}{\partial r^2} \end{pmatrix} = K + D\Delta$$

with

$$K = \begin{pmatrix} B - 1 & A^2 \\ -B & -A^2 \end{pmatrix} \quad \text{and} \quad D = \begin{pmatrix} d & 0 \\ 0 & \theta d \end{pmatrix} .$$

Clearly, the operator $u \to \Delta u$ with $\frac{\partial u}{\partial r} = 0$ at 0 and 1 has eigenvalues $-n^2\pi^2$ $(n = 0,1,\ldots)$ with corresponding eigenfunctions $\cos n\pi x$. Let $\varphi = \Sigma \begin{pmatrix} a_n \\ b_n \end{pmatrix} \cos n\pi x$ be an eigenfunction for L with eigenvalue λ. Now,

$$(K + D\Delta)\ (\Sigma \begin{pmatrix} a_n \\ b_n \end{pmatrix} \cos n\pi x)$$

$$= \Sigma\, K \begin{pmatrix} a_n \\ b_n \end{pmatrix} + D \begin{pmatrix} a_n \\ b_n \end{pmatrix} (-n^2 \pi^2) \cos n\pi x$$

$$= \lambda\, \Sigma \begin{pmatrix} a_n \\ b_n \end{pmatrix} \cos n\pi x .$$

Hence, $(K - \pi^2 n^2 D) \begin{pmatrix} a_n \\ b_n \end{pmatrix} = \lambda \begin{pmatrix} a_n \\ b_n \end{pmatrix} \quad (n = 0,1,\ldots)$.

It follows from this, that the eigenvalues of L are given by the eigenvalues of $K - \pi^2 n^2 D$ for $n = 0,1,2,\ldots$ Now, the trace of $K - \pi^2 n^2 D$ is

$$\mathrm{tr}(K - \pi^2 n^2 D) = B - 1 - A^2 - n^2 \pi^2 d(1 + \theta) .$$

For the Hopf bifurcation to occur at a given value of B, there must be a pure imaginary, conjugate pair of eigenvalues, i.e., the trace must vanish. The only possible critical values of B are therefore $B_m = 1 + A^2 + m^2 \pi^2 d(1 + \theta)$, $(m = 0,1,\ldots)$. At $B = B_m$, $\mathrm{tr}(K - \pi^2 n^2 D) = (m^2 - n^2)\pi^2 d(1 + \theta)$, so if $m > 0$. $\mathrm{tr}(K - \pi^2 n^2 D) > 0$ for all $0 \leq n < m$ and the operator L has at least $2m$ eigenvalues with positive real parts. The only value of B at which the Hopf hypotheses may be satisfied is therefore $B = B_0 = 1 + A^2$. Near B_0, the complex conjugate pair $\lambda_1 = \bar{\lambda}_2 = \alpha + i\omega$ is given by

$$\alpha = \frac{1}{2}[B - (1 + A^2)],\ \omega^2 = A^2 - \alpha^2 ,$$

while the remaining eigenvalues λ must satisfy

$$\lambda^2 - \lambda\, \mathrm{tr}(K - \pi^2 n^2 D) + \det(K - \pi^2 n^2 D) = 0 \quad \text{for some } n = 1,2,\ldots .$$

The Hopf hypotheses on the eigenvalues are fulfilled at $B = B_0$ if and only if $\det(K - \pi^2 n^2 D) > 0$ uniformly for all $n \geq 1$ and B in some neighborhood of B_0, which is true if and only if

$$\min_{n\geq 1} (\pi^4 n^4 \theta d^2 + \pi^2 n^2 A^2 d(1 - \theta) + A^2) > 0 .$$

We assume in the following that this condition is satisfied. If $\theta \leq 1$, this condition clearly holds.

To determine the stability of the bifurcating periodic solutions, we need to know the restriction of the system to its center manifold at $B = 1 + A^2$. Denote by L^* the operator

$$\begin{pmatrix} u \\ v \end{pmatrix} \rightarrow \begin{pmatrix} (B - 1) + D \dfrac{\partial^2}{\partial r^2} & -B \\ A^2 & -A^2 + \theta D \dfrac{\partial^2}{\partial r^2} \end{pmatrix} \begin{pmatrix} u \\ v \end{pmatrix}$$

with domain

$$\{(u,v) \in H^2[0,1] \times H^2[0,1] \mid \frac{\partial u}{\partial r} = \frac{\partial v}{\partial r} = 0 \text{ at } 0,1\} .$$

Choose

$$q = \begin{pmatrix} 1 \\ -1 + \dfrac{i}{A} \end{pmatrix}, \quad q^* = \frac{1}{2} \begin{pmatrix} 1 + Ai \\ Ai \end{pmatrix} .$$

It is easy to see that $\langle L^* a,b \rangle = \langle a,Lb \rangle$ for any a in D_{L^*}, b in D_L, $Lq = iAq$, $L^* q^* = -iAq^*$, $\langle q^*,q \rangle = 1$ and $\langle q^*,\bar{q} \rangle = 0$. Here $\langle f,g \rangle = \int_\Omega \bar{f}^T g$ denotes the ordinary inner product in $L^2 \times L^2$. Write

$$\begin{pmatrix} u \\ v \end{pmatrix} = zq + \bar{z}\bar{q} + w \; ; \; z = \langle q^*, \begin{pmatrix} u \\ v \end{pmatrix} \rangle .$$

Thus,

$$\begin{cases} u = z + \overline{z} + w_1 \\ v = z(-1 + \frac{i}{A}) + \overline{z}(-1 - \frac{i}{A}) + w_2 \end{cases}$$

Our system in z , w coordinates becomes:

$$\begin{cases} \frac{dz}{dt} = iAz + \langle q^*, f\rangle \\ \frac{dw}{dt} = Lw + [f - \langle q^*, f\rangle\, q - \langle \overline{q}^*, f\rangle\, \overline{q}] = Lw + H \end{cases}$$

with $f = \binom{h}{-h}$.

Simple computations show that

$$\langle q^*, f\rangle = \frac{h}{2} \quad \text{and} \quad H = 0 \; .$$

Write $w = \frac{w_{20}}{2} z^2 + w_{11} z\,\overline{z} + \frac{w_{02}}{2}\overline{z}^2 + O(|z|^3)$ for the equation of the center manifold. We get

$$\begin{cases} (2iA - L)w_{20} = 0 \; , \\ (-L)w_{11} = 0 \; , \\ w_{02} = \overline{w}_{20} \end{cases}$$

or $w_{20} = w_{02} = w_{11} = 0$.

Consequently, the equation on the center manifold in z, $\overline{z}$ coordinates now is:

$$\begin{aligned} \frac{dz}{dt} &= iAz + \frac{1}{2}h \, , \\ &= iAz + \frac{1}{2}[\frac{B}{A}u^2 + 2Auv + u^2 v] \; , \\ &= iAz + \frac{2(A^2+1)}{A}(\mathrm{Re}\, z)^2 + 4\,\mathrm{Re}[z(-1+i/A)][A\,\mathrm{Re}\, z + (\mathrm{Re}\, z)^2] + O(|z|^4) \; . \end{aligned}$$

Thus,

$$g_{20} = (\frac{1}{A} - A) + 2i ,$$

$$g_{11} = (\frac{1}{A} - A) ,$$

$$g_{02} = (\frac{1}{A} - A) - 2i ,$$

$$\text{and} \quad g_{21} = -3 + \frac{i}{A} .$$

Since $\text{Re } c_1(0) = \text{Re}\{ \frac{i}{2A} g_{20}g_{11} + \frac{g_{21}}{2}\}$, we obtain

$\text{Re } c_1(0) = -[\frac{1}{A^2} + \frac{1}{2}] < 0$. Therefore, we conclude that asymptotically orbitally stable periodic solutions bifurcate from the stationary solution $(A, B/A)$ as B passes through $1 + A^2$.

One should observe that the reaction equation for u,v without the diffusion term $D\Delta$ in $z, \bar{z}$ coordinates is the same (up to third order) as the restriction of our system on its center manifold. Thus, by the same reasoning for the Brusselator without diffusion, a family of periodic solutions are found, and they are space independent. In fact, we already made such a computation in Example 3 of Chapter 2, where the coordinate system is different from the present one by a factor of 2. These periodic solutions are called bulk oscillations of the Brusselator. By the uniqueness of the bifurcating periodic solutions for Hopf bifurcations, these bulk oscillations have to agree with the periodic solutions of the Brusselator with diffusion and no flux boundary condition just found through use of the center manifold theorem. Consequently, the bulk oscillations branching from the stationary state in the Brusselator are actually stable in the whole function space, a result which we promised in Example 3 of Chapter 2. In [7], Auchmuty and Nicholis obtained these same bulk oscillations; see also Boa and Cohen [11].

We next consider the Brusselator on a circular disk of adius a with no-flux boundary condition. The system to be onsidered is

$$\frac{\partial}{\partial t}\binom{u}{v} = (K + D\Delta)\binom{u}{v} + \begin{pmatrix} h(u,v) \\ -h(u,v) \end{pmatrix} ,$$

here now

$$\Delta = \frac{1}{r}\frac{\partial}{\partial r}\left(r\frac{\partial}{\partial r}\right) + \frac{1}{r^2}\frac{\partial^2}{\partial \phi^2} ,$$

$\leq r < a$ and $0 \leq \phi < 2\pi$, u and v are periodic in ϕ with eriod 2π , nonsingular at $(r,\phi) = (0,\phi)$, and $\partial u/\partial r = \partial v/\partial r = 0$ $= a$ and $0 \leq \phi < 2\pi$.

The eigenvalue problem

$$\Delta w = \lambda w \qquad (0 \leq r < a,\ 0 \leq \phi < 2\pi) ,$$

smooth, and $\partial w/\partial r = 0$ for $r = a,\ 0 \leq \phi < 2\pi$ has double igenvalues $-(\lambda_m^n)^2$ $(m = 0,\ n = 0,\ 1;\ m = 1,2,\ldots,\ n = 0,1,\ldots)$, efined by

$$J_n'(\lambda_m^n a) = 0 ,$$

nd corresponding eigenfunctions.

$$J_n(\lambda_m^n r)\cos n\phi ,\ J_n(\lambda_m^n r)\sin n\phi ,$$

here J_n is the usual Bessel function.

Note that $\lambda_0^n = 0$ $(n = 2,3,\ldots)$, and the corresponding eigenfunctions" are 0 .) It is known that for each n the λ_m^n re positive, except that $\lambda_0^0 = 0$, and strictly increase to ∞ with m .

As in the case of one space dimension considered above, it ollows that the eigenvalues of $L = K + D\Delta$ are given by the igenvalues of $K - (\lambda_m^n)^2 D$ $(m = 0,\ n = 0,\ 1;\ m = 1,2,\ldots,\ n = 0,1,\ldots)$.

For Hopf bifurcation with respect to B to occur

$$\operatorname{tr}(K - (\lambda_m^n)^2 D) = B - 1 - A^2 - (\lambda_m^n)^2 d(1 + \theta)$$

must vanish and $\det(K - (\lambda_m^n)^2 D)$ must be positive at that value of B. Now, as we expect, $\lambda_0^0 = 0$, the corresponding eigenfunction is 1 (this eigenvalue of Δ is simple); and the Hopf spectral hypotheses are met for $0 < \theta < 1$ if $B_c = 1 + A^2$, since

$$\lambda_m^n > 0 \quad (m = 0,\ n = 1;\ m = 1,2,\ldots,\ n = 0,1,\ldots)$$

and

$$\min[(\lambda_m^n)^4 \theta d^2 + (\lambda_m^n)^2 A^2 (1 - \theta)d + A^2] > 0 ,$$

where the minimum is taken over the allowed values of m and n. The remainder of the analysis is essentially the same is in the case of one space dimension; μ_2 and β_2 are exactly the same, and the bifurcating periodic solutions are stable bulk oscillations.

. THE BRUSSELATOR WITH DIFFUSION AND FIXED BOUNDARY CONDITIONS

Hopf bifurcation also occurs in the Brusselator on the unit nterval with Dirichlet (fixed) boundary conditions. The system s given by

$$\begin{cases} \dfrac{\partial X}{\partial t} = A - (B+1)X + X^2 Y + d\,\dfrac{\partial^2 X}{\partial r^2} \\[2ex] \dfrac{\partial Y}{\partial t} = BX - X^2 Y + \theta d\,\dfrac{\partial^2 Y}{\partial r^2} \end{cases}$$

ith boundary conditions

$$\begin{cases} X(0,t) = X(1,t) = A \\ Y(0,t) = Y(1,t) = B/A\ . \end{cases}$$

ere, r belongs to $[0,1]$, and X,Y are real-valued functions f two variables t,r . A, B, d, θ are positive numbers, and will be treated as the bifurcation parameter. Set

$$\begin{cases} x = X - A \\ y = Y - B/A\ . \end{cases}$$

he system becomes:

$$\begin{cases} \dfrac{\partial x}{\partial t} = (B-1)\,x + A^2 y + d\,\dfrac{\partial^2 x}{\partial r^2} + h(x,y) \\[2ex] \dfrac{\partial y}{\partial t} = -Bx - A^2 y + \theta d\,\dfrac{\partial^2 y}{\partial r^2} - h(x,y) \end{cases}$$

ith zero Dirichlet data on the boundary:

$$\begin{cases} x(0,t) = x(1,t) = 0 \\ y(0,t) = y(1,t) = 0\ . \end{cases}$$

ere $h(x,y) = \frac{B}{A}x^2 + 2Axy + x^2 y$.

From Example (c) in Section 5.2, we know that the Brusselator in x,y coordinates with zero Dirichlet data defines a smooth local semiflow F_t on $Y = \{(u,v) \mid u,v \in D_A$ and u,v are real$\}$. Here $D_A = \{u \in H^2[0,1] \mid u = 0$ at $0,1\}$, and $D_X F_t(0)$ is a compact semigroup with infinitesimal generator $\mathcal{L} = K + D\Delta$,

$$\text{where } K = \begin{pmatrix} B-1 & A^2 \\ -B & -A^2 \end{pmatrix} \text{ and } D = \begin{pmatrix} d & 0 \\ 0 & \theta d \end{pmatrix}.$$

The operator $u \to \frac{\partial^2 u}{\partial r^2}$ with $u \in D_A$ has eigenvalues $-n^2\pi^2$ $(n = 1,2,\ldots)$ and corresponding eigenfunctions $\sin n\pi x$. Let $\psi = \Sigma \begin{pmatrix} a_n \\ b_n \end{pmatrix} \sin n\pi x$ be an eigenfunction for $\mathcal{L}$ with eigenvalue λ. Then

$$(K + D\Delta)\psi = \Sigma\left[K\begin{pmatrix} a_n \\ b_n \end{pmatrix} + D\begin{pmatrix} a_n \\ b_n \end{pmatrix}(-n^2\pi^2)\right] \sin n\pi x,$$

$$= \lambda \Sigma \begin{pmatrix} a_n \\ b_n \end{pmatrix} \sin n\pi x.$$

Thus, $(K - n^2\pi^2 D)\begin{pmatrix} a_n \\ b_n \end{pmatrix} = \lambda \begin{pmatrix} a_n \\ b_n \end{pmatrix}$ $(n = 1,2,\ldots)$. Therefore, the eigenvalues of $\mathcal{L}$ are given by the eigenvalues of $K - n^2\pi^2 D$ $(n = 1,2,\ldots)$, which in turn satisfy

$$\lambda^2 - \lambda \operatorname{trace}(K - n^2\pi^2 D) + \det(K - n^2\pi^2 D) = 0,$$

where

$$\operatorname{trace}(K - n^2\pi^2 D) = B - B_n^c,$$

$$\det(K - n^2\pi^2 D) = n^2\pi^2\theta d(B_n^r - B),$$

$$B_n^c = 1 + A^2 + n^2\pi^2 d(1 + \theta), \quad \text{and}$$

$$B_n^r = (1 + n^2\pi^2 d)(1 + A^2/n^2\pi^2\theta d) \quad (n = 1,2,\ldots) .$$

Also let

$$\mathrm{disc}_n(B) = \left(\frac{B - B_n^c}{2}\right)^2 + n^2\pi^2\theta d(B - B_n^r) , \quad (n = 1,2,\ldots) .$$

For a Hopf bifurcation to occur at some value of B, there must be a pair of eigenvalues $\lambda = \pm i\omega_0$, with $\omega_0 > 0$. Thus for one of the matrices $K - n^2\pi^2 D$, the trace must vanish and the determinant must be positive, i.e. for some $n = m$, $B = B_m^c$ and $\omega_0^2 = m^2\pi^2\theta d\ (B_m^r - B_m^c) > 0$. If $m > 1$, then at $B = B_m^c$

$$\mathrm{trace}\ (K - n^2\pi^2 D) = (m^2 - n^2)\pi^2 d(1 + \theta)$$

is positive for all $1 \le n < m$ and the operator $\mathcal{L}$ has at least $m - 1$ eigenvalues with positive real parts. The only value of B at which the Hopf hypotheses may be satisfied is therefore

$$B = B_1^c = 1 + A^2 + \pi^2 d(1 + \theta) ,$$

and we must further assume that

$$\omega_0^2 \equiv -\mathrm{disc}_1(B_1^c) = \pi^2\theta d(B_1^r - B_1^c) > 0 .$$

Assume finally that at B_1^c

$$\det(K - n^2\pi^2 D) = n^2\pi^2\theta d\ (B_n^r - B_1^c) > 0$$

for $n \ge 2$. With these assumptions, it is straightforward (see Exercises 1, 2 and 3) to construct an interval I

containing B_1^c and having the properties

i) for $B \in I$, $\mathcal{L}$ has a complex conjugate pair of eigenvalues $\alpha \pm i\omega$, where $\alpha = \frac{1}{2}(B - B_1^c)$, $\omega^2 = -\mathrm{disc}_1(B)$,

ii) the remaining eigenvalues λ obey $\mathrm{Re}\lambda \leq -\varepsilon$ for all $B \in I$, where $\varepsilon > 0$ is independent of B.

Thus the Hopf hypotheses on the eigenvalues are satisfied at B_1^c.

It is computationally advantageous to perform a change of variables at this stage. Note that for B near B_1^c, the eigenvector of $K - \pi^2 D$ corresponding to λ, is $v_1 = (1, (\alpha + i\omega + d\pi^2 - B + 1)/A^2)^T$. We let

$$P_1 = \begin{pmatrix} 1 & 0 \\ \gamma & \delta \end{pmatrix} = [(\mathrm{Re}v_1)\ (-\mathrm{Im}v_1)],$$

where $\gamma = (\alpha + d\pi^2 - B + 1)/A^2$, $\delta = -\omega/A^2$, and change variables according to

$$\begin{pmatrix} x \\ y \end{pmatrix} = P_1 \begin{pmatrix} u \\ v \end{pmatrix}.$$

The system becomes

$$\begin{pmatrix} \dot{u} \\ \dot{v} \end{pmatrix} = L\begin{pmatrix} u \\ v \end{pmatrix} + \begin{pmatrix} 1 \\ \beta \end{pmatrix} h(u,v),$$

where

$$L = \begin{pmatrix} \alpha + d(\pi^2 + \frac{\partial^2}{\partial r^2}) & -\omega \\ \omega + (\theta - 1)\frac{\gamma d}{\delta}(\pi^2 + \frac{\partial^2}{\partial r^2}) & \alpha + \theta d(\pi^2 + \frac{\partial^2}{\partial r^2}) \end{pmatrix}$$

with the domain

$$D_L = \{(u,v) \mid u,v \in H^2[0,1] \text{ and } u = v = 0 \text{ at } 0,1\}$$

n the above

$$\beta = (\alpha + d\pi^2 - B + 1 + A^2)/\omega ,$$

$$\sigma = B/A + 2A\gamma , \text{ and}$$

$$h(u,v) = \sigma u^2 + 2A\delta\, uv + \gamma u^3 + \delta u^2 v .$$

Since we shall only compute μ_2, τ_2 and β_2, it suffices :o find the restriction of this system to the slice $B = B_1^c$ of :he center manifold, and so we set $B = B_1^c$ in the following.

Denote by L^* the operator

$$L^* = \begin{pmatrix} d(\pi^2 + \frac{\partial^2}{\partial r^2}) & \omega_0 + (\theta - 1)\frac{\gamma d}{\delta}(\pi^2 + \frac{\partial^2}{\partial r^2}) \\ -\omega_0 & \theta d\, (\pi^2 + \frac{\partial^2}{\partial r^2}) \end{pmatrix}$$

ith the same domain as L .

Choose

$$q = q^* = \begin{pmatrix} 1 \\ -i \end{pmatrix} \sin \pi r .$$

t is easy to check that $\langle L^*a,b\rangle = \langle a,Lb\rangle$ for all a in D_{L^*}, in D_L , and that $Lq = i\omega_0 q$, $L^*q^* = -i\omega_0 q^*$, $\langle q^*,q\rangle = 1$, $q^*,\bar{q}\rangle = 0$. Here,

$$\langle a,b\rangle = \int_0^1 \bar{a}^T b\, dr$$

enotes the usual inner product in $L^2 \times L^2$.

Write

$$\begin{pmatrix} u \\ v \end{pmatrix} = zq + \bar{z}\bar{q} + w; \; z = \langle q^*, \begin{pmatrix} u \\ v \end{pmatrix}\rangle .$$

Thus

$$u = (z + \bar{z}) \sin \pi r + w_1 = u_{||} + w_1 ,$$

$$v = i(\bar{z} - z) \sin \pi r + w_2 = v_{||} + w_2 .$$

Therefore

$$h(u,v) = \sigma(u_{||}^2 + 2u_{||} w_1) + 2A\delta(u_{||} v_{||} + u_{||} w_2 + v_{||} w_1)$$
$$+ \gamma u_{||}^3 + \delta u_{||}^2 v_{||} + O(|z|^4) ,$$

where we have assumed that $w = O(|z|^2)$ and have retained only the terms necessary to compute $c_1(0)$.

The integrals

$$\int_0^1 \sin \pi r \, dr = \frac{2}{\pi} , \quad \int_0^1 \sin^2 \pi r \, dr = \frac{1}{2} ,$$

$$\int_0^1 \sin^3 \pi r \, dr = \frac{4}{3\pi} , \quad \int_0^1 \sin^4 \pi r \, dr = \frac{3}{8} ,$$

are used below. One computes

$$\langle q^*, f \rangle = \langle q^*, \begin{pmatrix} h \\ \beta h \end{pmatrix} \rangle = (1 + i\beta) \int_0^1 h \sin \pi r \, dr ,$$

where

$$\int_0^1 h \sin \pi r \, dr = \sigma\{(z + \bar{z})^2 \frac{4}{3\pi} + 2(z + \bar{z})\hat{w}_1\}$$
$$+ 2A\delta\{i(\bar{z}^2 - z^2) \frac{4}{3\pi} + (z + \bar{z})\hat{w}_2 + i(\bar{z} - z)\hat{w}_1\}$$
$$+ \gamma(z + \bar{z})^3 \frac{3}{8} + \delta i(z + \bar{z})(\bar{z}^2 - z^2) \frac{3}{8} + O(|z|^4) .$$

Here $\begin{pmatrix} \hat{w}_1 \\ \hat{w}_2 \end{pmatrix} = \hat{w} = \int_0^1 w \sin^2 \pi r \, dr$.

The system in z, w coordinates is then

$$\frac{dz}{dt} = i\omega_0 z + \langle q^*, f \rangle$$

$$\frac{dw}{dt} = Lw + f - 2\text{Re}(\langle q^*, f \rangle q)$$

$$= Lw + \begin{pmatrix} 1 \\ \beta \end{pmatrix} h_\perp ,$$

where

$$h_\perp = h - 2 \sin \pi r \int_0^1 h \sin \pi r \, dr$$

$$= [\sigma(z + \bar{z})^2 + 2A\delta i(\bar{z}^2 - z^2)](\sin^2 \pi r)_\perp + O(|z|^3) ,$$

$$(\sin^2 \pi r)_\perp = \sin^2 \pi r - \frac{8}{3\pi} \sin \pi r = \Sigma_3^\infty I_n \sin n\pi r ,$$

and

$$I_n = 2 \int_0^1 \sin^2 \pi r \sin n\pi r \, dr = \begin{cases} 0 & \text{if } n \text{ is even} \\ \dfrac{8}{\pi n(4 - n^2)} & \text{if } n \text{ is odd.} \end{cases}$$

Now, if $w = \frac{w_{20}}{2} z^2 + w_{11} z\bar{z} + \frac{w_{02}}{2} \bar{z}^2 + \ldots$ represents the center manifold, $w_{02} = \bar{w}_{20}$ and w_{20} , w_{11} must solve the linear, two-point boundary value problems

$$(2i\omega_0 - L) w_{20} = 2(\sigma - 2A\delta i) \begin{pmatrix} 1 \\ \beta \end{pmatrix} (\sin^2 \pi r)_\perp ,$$

$$-Lw_{11} = 2\sigma \begin{pmatrix} 1 \\ \beta \end{pmatrix} (\sin^2 \pi r)_\perp$$

with $w_{20} = w_{11} = 0$ at $r = 0,1$. Although these problems may be solved for w_{20}, w_{11} in terms of closed form expressions, the solutions are more easily described in terms of Fourier series

$$w_{20} = -2(\sigma - 2A\delta i)\Sigma_3^\infty I_n \sin n\pi r (L_n - 2i\omega_0)^{-1}\begin{pmatrix}1\\ \beta\end{pmatrix},$$

$$w_{11} = -2\sigma\Sigma_3^\infty I_n \sin n\pi r L_n^{-1}\begin{pmatrix}1\\ \beta\end{pmatrix},$$

where

$$L_n = \begin{pmatrix} d\pi^2(1 - n^2) & -\omega_0 \\ \omega_0 + (\theta - 1)\frac{\gamma}{\delta} d\pi^2(1 - n^2) & \theta d\pi^2(1 - n^2) \end{pmatrix}.$$

Now

$$\hat{w} = \int_0^1 w \sin^2 \pi r\, dr = \hat{w}_{11} z\bar{z} + \mathrm{Re}\,(\hat{w}_{20} z^2) + O(|z|^3),$$

where

$$\hat{w}_{11} = \int_0^1 w_{11} \sin^2 \pi r\, dr = -\sigma\, \Sigma_3^\infty I_n^2 L_n^{-1}\begin{pmatrix}1\\ \beta\end{pmatrix},$$

$$\hat{w}_{20} = \int_0^1 w_{20} \sin^2 \pi r\, dr = -(\sigma - 2A\delta i)\, \Sigma_3^\infty I_n^2 (L_n - 2i\omega_0)^{-1}\begin{pmatrix}1\\ \beta\end{pmatrix}.$$

Restricted to the center manifold, the system therefore has the form

$$\frac{dz}{dt} = i\omega_0 z + \sum_{2\le i+j\le 3} \frac{g_{ij}}{i!j!} z^i \bar{z}^j + O(|z|^4),$$

where the coefficients g_{ij} arise in the expansion of

$$\langle q^*, f\rangle = (1 + i\beta)\int_0^1 h \sin \pi r \, dr \; ;$$

thus $g_{ij} = (1 + i\beta)h_{ij}$, $2 \le i + j \le 3$, where

$$\sum_{2\le i+j\le 3} \frac{h_{ij}}{i!j!} z^i \bar{z}^j = \int_0^1 h \sin \pi r \, dr + 0\,(|z|^4) \, ,$$

and

$$h_{11} = \frac{8\sigma}{3\pi} \, , \quad h_{20} = h_{11} - \frac{16A\delta i}{3\pi} \, , \quad h_{02} = \bar{h}_{20} \, ,$$

$$\frac{h_{21}}{2} = 2(\sigma - A\delta i)\,\hat{w}^1_{11} + 2A\delta\hat{w}^2_{11}$$

$$+ (\sigma + A\delta i)\,\hat{w}^1_{20} + A\delta\hat{w}^2_{20} + \frac{3}{8}\,(3\gamma - i\delta) \, .$$

Finally,

$$c_1(0) = \frac{i}{2\omega_0}\,\{g_{20}g_{11} - 2|g_{11}|^2 - \frac{1}{3}\,|g_{02}|^2\} + \frac{g_{21}}{2}$$

$$= \frac{i}{2\omega_0}\,\{(1 + i\beta)^2 h_{20}h_{11} - (1 + \beta^2)(2|h_{11}|^2 + \tfrac{1}{3}|h_{02}|^2)\}$$

$$+ (1 + i\beta)\,\frac{h_{21}}{2} \, ,$$

where the h_{ij}'s are given above.

The parameters μ_2, τ_2, β_2 are then $\mu_2 = -\mathrm{Re}\, c_1(0)/\alpha'(0)$, $\tau_2 \equiv -[\mathrm{Im}\, c_1(0) + \mu_2\, \omega'(0)]/\omega_0$, $\beta_2 = 2\,\mathrm{Re}\, c_1(0)$, where $\alpha'(0) = 1/2$ and $\omega'(0) = -d\theta\pi^2/2\omega_0$.

The periodic solutions themselves are approximated as

$$x(t,r) = u(t,r)$$

$$y(t,r) = \gamma u(t,r) + \delta v(t,r)$$

$$\binom{u}{v} = zq + \bar{z}\bar{q} + O(|z|^2)$$

$$= \begin{pmatrix} z + \bar{z} \\ i(\bar{z} - z) \end{pmatrix} \sin \pi r + O(|z|^2) ,$$

$$z(t) = \varepsilon e^{2\pi i t/T} + O(\varepsilon^2) ,$$

where ε satisfies

$$\mu_2 \varepsilon^2 + O(\varepsilon^4) = B - B_1^c .$$

To measure the amplitude of these periodic solutions, we compute

$$\frac{1}{T}\int_0^T \int_0^1 x^2(t,r) + y^2(t,r)\,dr]dt = (1 + \gamma^2 + \delta^2)\varepsilon^2 + O(\varepsilon^3)$$

$$= [(B - B_1^c)/\mu_2](1 + \gamma^2 + \delta^2) + O(\varepsilon^3) ,$$

this last equality being valid in the case $\mu_2 \neq 0$. To compare the present results with those of Chapter 3, Example 5, we now define

$$\bar{\mu}_2 = \mu_2/(1 + \gamma^2 + \delta^2) ,$$

$$\bar{\tau}_2 = \tau_2/(1 + \gamma^2 + \delta^2) ,$$

$$\bar{\beta}_2 = \beta_2/(1 + \gamma^2 + \delta^2) .$$

For the special set of parameter values $A = 1$, $d = .1$, $\theta = .5$ we find $B_1^c = 3.48044066$, $\omega_0 = 1.11801498$, $\gamma = -1.49348022$, $\delta = -1.11801498$, $\beta = -.44138963$, $\sigma = .49348022$,
$\hat{w}_{11} = (.00166124,\ .00108552)^T$,
$\hat{w}_{20} = (.00317821 + .00708898i,\ .00448136 + .00167780i)^T$,

$h_{11} = .41887902$, $h_{20} = .41887902 + 1.89800118i$, $h_{02} = \bar{h}_{20}$,
$h_{21} = -3.35293837 + .84207874i$, $\mu_2 = 3.41528048$,
$\tau_2 = .06804667$, $\beta_2 = -\mu_2$, $\bar{\mu}_2 = .76226442$, $\bar{\tau}_2 = .01518750$,
and $\bar{\beta}_2 = -\bar{\mu}_2$. These values for $\bar{\mu}_2$, $\bar{\tau}_2$ and $\bar{\beta}_2$ agree quite well with those in Chapter 3, Example 5.

Remark: In calculations as complicated as the present, it is best to assume that one's work contains errors, at least until much supporting evidence is available.

The formula for $c_1(0)$ is in general a transcendental function of the parameters d, θ, and A , which we shall evaluate on a 3-dimensional grid of points. Before discussing these results, however, we note that for fixed A, θ, $c_1(0)$ is continuous in d at 0 . But for $d = 0$, all of the matrices L_n coincide and the expression for $c_1(0)$ simplifies. We find

$$L_n = \begin{pmatrix} 0 & -\omega_0 \\ \omega_0 & 0 \end{pmatrix} \qquad (n = 1,2,\ldots) ,$$

$\alpha_1 = 1 + A^2$, $\omega_0 = A$, $\beta = 0$, $\gamma = -1$, $\delta = -1/A$ and $\sigma = (1 - A^2)/A$. Thus

$$\hat{w}_{11} = \frac{\sigma}{A} \begin{pmatrix} 0 \\ 1 \end{pmatrix} \Sigma_3^\infty I_n^2 ,$$

and

$$\hat{w}_{20} = -\frac{1}{3A} (\sigma + 2i) \begin{pmatrix} 2i \\ 1 \end{pmatrix} \Sigma_3^\infty I_n^2 ,$$

where $\Sigma_3^\infty I_n^2 = 2 \int_0^1 [(\sin^2 \pi r)_\perp]^2 \, dr = \frac{3}{4} - (\frac{8}{3\pi})^2$.

After some simplification, we obtain

$$h_{11} = \frac{8\sigma}{3\pi}, \quad h_{20} = h_{11} + \frac{16}{3\pi} i, \quad h_{02} = \bar{h}_{20},$$

$$\frac{h_{21}}{2} = -\frac{1}{A} [\sigma + \frac{2}{3} i (1 + \sigma^2)][\frac{3}{4} - (\frac{8}{3\pi})^2] - \frac{9}{8} + \frac{3i}{8A},$$

and hence

$$c_1(0) = -\frac{3}{4} (\frac{1}{A^2} + \frac{1}{2}) - i (\frac{1}{2A^3} - \frac{7}{8A} + \frac{A}{2}).$$

This expression is precisely 3 times the expression for $c_1(0)$ in the case of reaction alone (Chapter 2). This observation turns out to be a special case of a more general result, see Remark 1 below. In the present situation, we have established the following: if θ and A are fixed, then for all sufficiently small positive d, the first bifurcation that occurs from the constant solution $(X, Y) = (A, B/A)$ as B is increased from 0 is a Hopf bifurcation, occurs at $B = B_1^c$ and is supercritical, giving rise to asymptotically, orbitally stable, small amplitude periodic solutions.

To investigate the Hopf bifurcation for more general parameter values d, θ, A, we evaluated $c_1(0; d, \theta, A)$ for $A = 1/2, 1, 2, 4, 8$, $d = .2i$ $(0 \le i \le 5)$ and $\theta = .1j$ $(0 \le j \le 15)$. Table 5.1 was constructed from the results. The symbol '+' indicates supercritical bifurcation to stable periodic solutions, and the symbol '-' indicates subcritical bifurcation to unstable periodic solutions. A blank entry indicates that for this combination of parameters d, θ, A, the first bifurcation that occurs as B is increased from 0 is not a Hopf bifurcation.

We note that both supercritical and subcritical bifurcations do indeed occur. The subcritical bifurcations might indicate the presence of an additional stable stationary state or of a stable, larger amplitude periodic solution for $B = B_1$, $B < B_1$. It would not be difficult to investigate this situation

Table 5.1. Sign of $\mu_2(d,\theta,A)$ for the Brusselator with diffusion and Dirichlet boundary conditions

A = .25

d\θ	0					0.5					1.0		
.0	+	+	+	+	+	+	+	+	+	+	+	+	+
.1	+	-	-	-									
.2	+	-	-										
.3	+	-											
.4	+	-											
.5	+	-											
.6	+	-											
.7	+												
.8	+												
.9	+												
1.0	+												

A = 2.0

d\θ	0					0.5					1.0		
.0	+	+	+	+	+	+	+	+	+	+	+	+	+
.1	+	+	+	+	+	+	+	+	+	+	+	+	+
.2	+	+	+	+	+	+	+	+	+	+	-		
.3	+	+	+	+	+	+	+	+	+				
.4	+	+	+	+	+	+	+	+					
.5	+	+	+	+	+	+	+						
.6	+	+	+	+	+	+	+						
.7	+	+	+	+	+	+							
.8	+	+	+	+	+	+							
.9	+	+	+	+	+	-							
1.0	+	+	+	+	+								

A = .5

d\θ	0					0.5					1.0		
.0	+	+	+	+	+	+	+	+	+	+	+	+	+
.1	+	+	+	-	-	-							
.2	+	+	-	-									
.3	+	-	-										
.4	+	-	-										
.5	+	-	-										
.6	+	-	-										
.7	+	-											
.8	+	-											
.9	+	-											
1.0	+	-											

A = 4.0

d\θ	0					0.5					1.0		
.0	+	+	+	+	+	+	+	+	+	+	+	+	+
.1	+	+	+	+	+	+	+	+	-	-	-	-	-
.2	+	+	+	+	+	+	+	-	-	-	-	-	-
.3	+	+	+	+	+	+	+	-	-	-	-	-	
.4	+	+	+	+	+	+	+	-	-	-	-		
.5	+	+	+	+	+	+	+	-	-	-			
.6	+	+	+	+	+	+	+	-	-				
.7	+	+	+	+	+	+	+	-	-				
.8	+	+	+	+	+	+	+	-	-				
.9	+	+	+	+	+	+	+	-					
1.0	+	+	+	+	+	+	+	-					

A = 1.0

d\θ	0					0.5					1.0		
.0	+	+	+	+	+	+	+	+	+	+	+	+	+
.1	+	+	+	+	+	+	+	+	+	+	+		
.2	+	+	+	+	+	+	-						
.3	+	+	+	+	+	-							
.4	+	+	+	+	-								
.5	+	+	+	-	-								
.6	+	+	-	-									
.7	+	+	-	-									
.8	+	+	-	-									
.9	+	+	-	-									
1.0	+	+	-										

A = 8.0

d\θ	0					0.5					1.0		
.0	+	+	+	+	+	+	+	+	+	+	+	+	+
.1	+	+	+	+	+	-	-	-	-	-	-	-	-
.2	+	+	+	+	+	-	-	-	-	-	-	-	
.3	+	+	+	+	+	-	-	-	-	-	-	-	
.4	+	+	+	+	+	-	-	-	-	-	-	-	
.5	+	+	+	+	+	-	-	-	-	-	-	-	
.6	+	+	+	+	+	-	-	-	-	-	-		
.7	+	+	+	+	+	-	-	-	-	-	-		
.8	+	+	+	+	+	-	-	-	-	-	-		
.9	+	+	+	+	+	-	-	-	-	-			
1.0	+	+	+	+	+	-	-	-	-	-			

numerically; appropriate values for d, θ, A may be taken from Table 5.1.

Remark 1.

Let

$$\frac{dX}{dt} = (K_\mu + \delta D_\mu \Delta)X + f_\mu(X)$$

represent a reaction-diffusion equation on a smooth domain Ω in $\mathbb{R}^n$, $n \leq 3$ with either Dirichlet or Neumann boundary conditions, as in Section 2 c) or d). We shall show that when Hopf bifurcation from $X = 0$ occurs at $\mu = \mu(\delta)$, with $\mu(\delta) \to 0$ as $\delta \to 0+$, then the $c_1(\delta)$ of the system is of the form $M\, c_1^R(0) + O(\delta)$, where $M > 0$ and $c_1^R(0)$ is the $c_1(0)$ for the underlying reaction equation alone. Hence, if $c_1^R(0)$ indicates that stable periodic solutions bifurcate for the reaction system, then for all sufficiently small positive δ, the bifurcation for the reaction-diffusion system is also to stable periodic solutions.

Let us sketch the computations which show that $c_1 = Mc_1^R(0) + O(\delta)$. Write

$$f_0(X) = \frac{1}{2} Q(X,X) + \frac{1}{6} C(X,X,X) + O(|X|^4),$$

where Q and C are symmetric multilinear forms. For simplicity in notation, set $K_0 = K$, and $Q_{XY} = Q(X,Y)$.

By our assumptions, the reaction equations have a Hopf bifurcation at $\mu = 0$. Thus, there exist a, a^* in $\mathbb{C}^n$ such that $Ka = i\omega_0 a$, $K^*a^* = -i\omega_0 a^*$, $(a^*,a) = 1$, and $(a^*,\bar{a}) = 0$. Here, $(a,b) = \sum_{i=1}^{n} \bar{a}_i b_i$ for $a = (a_1,\ldots,a_n)$, $b = (b_1,\ldots,b_n)$. To each X in $\mathbb{R}^n$, let $z = (a^*,X)$. Thus, $X = az + \bar{a}\bar{z} + w$ with $(a^*,w) = (\bar{a}^*,w) = 0$. Write

$$w = \frac{w_{20}}{2} z^2 + w_{11} z\bar{z} + \frac{w_{02}}{2} \bar{z}^2 + \ldots$$

for the equation of a center manifold. It is not hard to see that

$$w_{20} = (2i\omega_0 - K)^{-1} (Q_{aa} - (a^*,Q_{aa})a - (\bar{a}^*,Q_{aa})\bar{a})$$

$$w_{11} = -K^{-1}(Q_{a\bar{a}} - (a^*,Q_{a\bar{a}})\, a - (\bar{a}^*,\, Q_{a\bar{a}})\bar{a}) \; .$$

The reaction equations restricted to their center manifold are:

$$\frac{dz}{dt} = i\omega_0 z + (a^*,f_0) = i\omega_0 z + \sum_{i,j=1}^{3} \frac{g_{ij} z^i \bar{z}^j}{i!j!} + O(|z|^4) \; ,$$

where

$$g_{20} = (a^*,Q_{aa})$$

$$g_{11} = (a^*,Q_{a\bar{a}}) \tag{5.5}$$

$$g_{02} = (a^*,Q_{\bar{a}\bar{a}})$$

and

$$g_{21} = 2(a^*,Q(w_{11},a)) + (a^*,Q(w_{20},\bar{a})) + (a^*,C(a,a,\bar{a})) \; .$$

Recall that $c_1^R(0) = \frac{i}{2\omega_0}\{g_{20}g_{11} - 2|g_{11}|^2 - \frac{1}{3}|g_{02}|^2\} + \frac{g_{21}}{2}$.

By our assumptions, the reaction-diffusion equation has a Hopf bifurcation at $\mu = \mu(\delta)$ with $\mu(\delta) \to 0$ as $\delta \to 0+$. Thus, there exist C^n-valued functions on Ω in the form $q = a\phi + O(\delta)$, $q^* = a^*\psi + O(\delta)$ such that $Lq = i\omega_\mu q$. $L^*q^* = -i\omega_\mu q^*$, $(q^*,q) = 1$ and $(q^*,\bar{q}) = 0$. Here $L = K_\mu + \delta D\Delta$, $\mu = \mu(\delta)$, and ϕ,ψ are real-valued functions on Ω . In our situation, ϕ,ψ are eigenfunctions of Δ and $\psi = \lambda\phi$

for some constant $\lambda > 0$. As before, set $X = qz + \bar{q}\bar{z} + W$ with $z = \langle q^*, X\rangle$ and write

$$W = \frac{W_{20}}{2} z^2 + W_{11} z\bar{z} + \frac{W_{02}}{2} \bar{z}^2 + \ldots$$

for the equation of its center manifold. Then,

$$W_{20} = (2i\omega_0 - K_1)^{-1}\{\phi^2 Q_{aa} - \langle\psi,\phi^2\rangle\phi(a^*, Q_{aa})a - \langle\psi,\phi^2\rangle\phi(\bar{a}^*, Q_{aa})\bar{a}\} + O(\delta)$$

$$W_{11} = -K^{-1}\{\phi^2 Q_{a\bar{a}} - \langle\psi,\phi^2\rangle\phi(a^*, Q_{a\bar{a}})a - \langle\psi,\phi^2\rangle\phi(\bar{a}^*, Q_{a\bar{a}})\bar{a}\} + O(\delta) .$$

Therefore,

$$W_{20} = \phi^2 w_{20} + (\phi^2 - \langle\psi,\phi^2\rangle\phi)\left[\frac{g_{20}}{i\omega_0} a + \frac{\bar{g}_{02}}{3i\omega_0} \bar{a}\right] + O(\delta)$$

$$W_{11} = \phi^2 w_{11} + (\phi^2 - \langle\psi,\phi^2\rangle\phi)\left[\frac{g_{11}}{-i\omega_0} a + \frac{\bar{g}_{11}}{i\omega_0} \bar{a}\right] + O(\delta) . \tag{5.6}$$

Denote by $\frac{dz}{dt} = i\omega_\mu z + \sum_{i+j=1}^{3} \frac{G_{ij}}{i!j!} z^i \bar{z}^j + O(|z|^4)$ the restriction of the reaction-diffusion equation on its center manifold at $\mu = \mu(\delta)$. Thus,

$$G_{20} = \langle\psi,\phi^2\rangle g_{20} + O(\delta) ,$$

$$G_{11} = \langle\psi,\phi^2\rangle g_{11} + O(\delta) , \tag{5.7}$$

$$G_{02} = \langle\psi,\phi^2\rangle g_{02} + O(\delta) ,$$

and

$$G_{21} = 2\langle q^*, Q(W_{11}, q)\rangle + \langle q^*, Q(W_{20}, \bar{q})\rangle + \langle q^*, C(q,q,\bar{q})\rangle .$$

Using (5.5), (5.6), we obtain

$$G_{21} = \langle\psi,\phi(\phi^2 - \langle\psi,\phi^2\rangle\phi)\rangle\{-\frac{g_{11}}{i\omega_0}g_{20} + 2\frac{\bar{g}_{11}}{i\omega_0}g_{11} + \frac{\bar{g}_{02}}{3i\omega_0}g_{02}\}$$

$$+ \langle\psi,\phi^3\rangle\{2(a,Q(w_{11},a)) + (a^*,Q(w_{20},\bar{a}))+(a^*,C(a,a,\bar{a}))\}+O(\delta)$$

$$(5.8)$$

$$= \langle\psi,\phi^3\rangle\{\frac{i}{\omega_0}[g_{20}g_{11} - 2|g_{11}|^2 - \frac{1}{3}|g_{02}|^2] + g_{21}\}$$

$$- \langle\psi,\phi^2\rangle^2\frac{i}{\omega_0}[g_{20}g_{11} - 2|g_{11}|^2 - \frac{1}{3}|g_{02}|^2]\} + O(\delta) .$$

Consequently, by (5-7) and (5.8), we get

$$c_1 = \frac{i}{2\omega_\mu}\{G_{20}G_{11} - 2|G_{11}|^2 - \frac{1}{3}|G_{02}|^2\} + \frac{G_{21}}{2}$$

$$= \langle\psi,\phi^3\rangle\{\frac{i}{2\omega_0}[g_{20}g_{11} - 2|g_{11}|^2 - \frac{1}{3}|g_{02}|^2] + \frac{g_{21}}{2}\} + O(\delta)$$

$$= \langle\psi,\phi^3\rangle\, c_1^R(0) + O(\delta) .$$

In our situation, $M = \langle\psi,\phi^3\rangle = \langle\lambda\phi, \phi^3\rangle = \lambda\langle\phi^2,\phi^2\rangle > 0$.

Remark 2.

Auchmuty and Nicholis [7] also considered the Brusselator on the unit interval with fixed boundary conditions. Their computations were formal, but basically followed Hopf's original method. The most complicated terms in the lengthy expression determining the direction of bifurcation were omitted, so we have not checked their results.

6. EXERCISES

Exercise 1.

Gershgorin's lemma states: The eigenvalues of an arbitrary n by n matrix A all lie in the union of closed discs D_i , $i = 1,\ldots,n$, in the complex plane, constructed as follows. The center of D_i is the i'th diagonal element of A and the radius of D_i is the sum of the absolute values of the off-diagonal elements in the i'th row. Prove Gerschgorin's lemma.

Exercise 2.

Apply Gershgorin's lemma to the matrix $K - n^2\pi^2 D$ of Section 4, and establish the result: for fixed positive A, d, θ and for a fixed bounded interval I, there is an $n_0 > 0$ such that for any $B \in I$ and $n \geq n_0$, the eigenvalues λ of $K - n^2\pi^2 D$ satisfy $\text{Re}\,\lambda \leq -\varepsilon < 0$, for some $\varepsilon > 0$ independent of B,n.

Exercise 3.

Suppose that, at some point $B = B_m$ $(m \neq 0)$ the eigenvalues of $K - m^2\pi^2 D$ form a pure imaginary pair $\pm i\omega_0$, where $\omega_0 > 0$. Also suppose that, at $B = B_m$, the eigenvalues of $K - n^2\pi^2 D$ for $n \geq m$ all have negative real parts. Using continuity of the individual eigenvalues with respect to B and the result of Exercise 2 above, show that there is an interval I containing B_m such that for any $B \in I$, the eigenvalues of $K - m^2\pi^2 D$ form a complex conjugate pair while the eigenvalues λ of $K - n^2\pi^2 D$ for $n > m$ all satisfy $\text{Re}\,\lambda \leq -\varepsilon < 0$, where $\varepsilon > 0$ is independent of B,n.

Exercise 4.

Show that for a scalar parabolic equation Hopf bifurcation never occurs.

Exercise 5.

Try to carry out the Bifurcation Analysis for the Brusselator on a Disk with Dirichlet boundary condition and analyze the limit as the diffusion tends to 0. This is a lengthy calculation, which we have not carried out.

Exercise 6.

Develop a theory of stationary bifurcations (cf. Exercise 12 of Chapter 2) for reaction-diffusion systems, analogous to the Hopf bifurcation theory in the present Chapter. Apply the theory to analyze the stationary bifurcations which occur in the

russelator on the unit interval with Dirichlet boundary condition
ions.

xercise 7.

Suppose that one has access to a library of mathematical oftware, including routines for the solution of linear and non-inear two-point boundary value problems, linear ordinary differ-ntial eigenvalue problems, and quadrature.

a) Describe the sequence of mumerical procedures one would use o repeat the computation in Section 5 .

b) Suppose that the Brusselator on the unit interval with irichlet boundary conditions has a nonconstant stationary solu-ion, which loses stability in a Hopf bifurcation. How is the equence of computations in (a) changed?

c) Consider the Brusselator with Dirichlet boundary conditions n the unit disk instead of on the unit interval. What pieces of athematical software are needed to repeat the computations in (a) nd (b)?

Appendix A. The Center Manifold Theorem

Here, we present a theory of center manifolds that is sufficiently general to cover the ordinary, delay and partial differential equations examined in this book. For general references on invariant manifolds and center manifold theories, see Kelley [72], Hartman [43], Hirsch-Shub-Pugh [50], Marsden-McCracken [81], and Iooss [61].

Consider an abstract differential equation

$$x' = f(x) \equiv Lx + h(x) \tag{*}$$

with x in some open set U in a real Hilbert space H containing the origin. Here, L is the infinitesimal generator of a C^0 semigroup e^{Lt} and $h(\cdot)$ is a H-valued C^r $(r \geq 1)$ function on U such that $h(0) = 0$ and $Dh(0) = 0$. A submanifold M in U with $0 \in M$ is said to be <u>locally invariant</u>, if to each x in M the solution $\phi_t(x)$ of (*) with initial condition $\phi_0(x) = x$ remains in M for some interval $0 \leq t < \tau$, where $\tau = \tau(x) > 0$. The tangent space T_0M of M at 0 is invariant under the linear operator L.

If $H = R^n$, the equation (*) is an ordinary differential equation as in Chapters 2 and 3. If one takes $H = C[-r,0]$ and $L = L_\mu$ as in equation (2.1) of Chapter 4, the form (*) represents a delay equation. If $H = Y$ and $L = \tilde{A}$ as in examples c) or d) of Chapter 5, the form (*) reduces to a reaction-diffusion equation.

Assume now that L satisfies the following spectral conditions:

a) H is a direct sum of two closed invariant spaces V_s, V_c, where $\mathrm{Re}\,\sigma(L|V_s) < \alpha < 0$ for some α and V_c has finite dimension with $\mathrm{Re}\,\sigma(L|V_c) = 0$.

b) $\sigma(e^{Lt}) = e^{\sigma(L)t} \cup \{0\}$ for all $t > 0$.

For ordinary differential equations with $\sigma(L) \leq 0$, the spectral conditions are always met.

<u>Definition</u>. A locally invariant manifold M is called a <u>center manifold</u> if $T_0M = V_c$.

Note that for simplicity, we merely study the center manifold in the absence of an unstable part of L .

The following example shows that there may be more than one center manifold locally for an equation (*).

<u>Example</u>. Consider the system

$$x' = x^2$$

$$y' = -y$$

For any real number α , the set $M_\alpha = \{(x,y) \mid y = \alpha e^{1/x}, \ x < 0\} \cup \{(x,y) \mid x \geq 0, \ y = 0\}$ is a center manifold for the system.

<u>The Center Manifold Theorem</u>

(a) <u>Under the hypotheses above on</u> L <u>and</u> h , <u>the equation (*) possesses a</u> (C^{r-1}) <u>center manifold</u> M .

(b) M <u>is locally attractive in the following sense. There exists an open neighborhood</u> U <u>of</u> 0 <u>such that whenever</u> $\phi_t(x) \in U$ <u>for all</u> $t \geq 0$, $\phi_t(x)$ <u>approaches</u> M <u>as</u> $t \to \infty$.

<u>Remark</u> 1. In the case of ordinary differential equations $(H = \mathbb{R}^n)$, a C^r equation (*) does possess a C^r center manifold [43].

<u>Remark</u> 2. If the equation (*) is of class C^∞ , a C^∞ center manifold may fail to exist. Van Strien [105] shows that the C^∞ system

$$x' = -x^2 + \mu^2$$

$$y' = -y - (x^2 - \mu^2)$$

$$\mu' = 0$$

has no C^∞ center manifold.

<u>Remark</u> 3. If the system (*) is analytic, an analytic center manifold may fail to exist. However, there is at most one analytic center manifold. For instance, the analytic system

$$x' = -x^2$$

$$y' = -y + x^2$$

has no analytic center manifold. For if this system had an analytic center manifold it would be represented by the divergent series $y = \sum_{n=2}^{\infty} (n-1)!x^n$.

<u>Remark</u> 4. The existence of a center manifold for ordinary differential equations in a more general setting can be found in Kelley [72].

We shall now sketch a proof of the center manifold theorem stated above. We use a combination of the methods used by Hartman [43] and Marsden-McCracken [81]. Thus, the center manifold will be obtained as an invariant manifold of the time one map of the associated semiflow.

<u>Outline of proof</u>

Let (x_c, x_s) denote the coordinates on H defined by the decomposition $H = V_c \oplus V_s$, and denote the norm of H by $|\cdot|$. Then the equation (*) becomes

$$x_c' = Cx_c + Y(x_c, x_s)$$
$$x_s' = Ax_s + Z(x_c, x_s) \quad . \qquad (**)$$

Let δ obeying $-\alpha > \delta > 0$ be given. Since $Re[\sigma(A)] < \alpha < 0$ and $Re[\sigma(C)] = 0$, one can find a norm $\|\cdot\|$ on H and a constant $k > 0$ with

$$\frac{1}{k}|\cdot| \le \|\cdot\| \le k|\cdot| \, ,$$

and such that

$$\|e^{At}\| \le e^{\alpha t} \qquad (t > 0)$$

and

$$e^{-\delta t} \le \|e^{Ct}\| \le e^{\delta t} \qquad (t > 0) \, .$$

Denote by $\psi: \mathbb{R} \to \mathbb{R}$ a C^∞ function such that

$$\psi(x) = \begin{cases} 1 & \text{for } |x| \le 1/(4k) \, , \\ >0 & \text{for } 1/(4k) \le |x| \le 1/(2k) \, , \\ 0 & \text{for } |x| \ge 1/(2k) \, . \end{cases}$$

Let ε_1 be a positive number such that $Y(y,z)$ and $Z(y,z)$ are both defined for $|(y,z)| < \varepsilon_1$; and for any ε with $0 < \varepsilon < \varepsilon_1$ let

$$\bar{Y}(y,z) = \begin{cases} \frac{1}{\varepsilon}\psi(|(y,z)|) \, Y(\varepsilon y, \varepsilon z) & \text{if } |(y,z)| < 1/(2k) \\ 0 & \text{if } |(y,z)| \ge 1/(2k) \, , \end{cases}$$

and

$$\bar{Z}(y,z) = \begin{cases} \frac{1}{\varepsilon}\psi(|(y,z)|) \, Z(\varepsilon y, \varepsilon z) & \text{if } |(y,z)| < 1/(2k) \\ 0 & \text{if } |(y,z)| \ge 1/(2k) \, . \end{cases}$$

Then $\overline{Y}$ and $\overline{Z}$ are defined for all $(y,z) \in H$. Define $\lambda(\Phi,\Psi)$ by

$$\sup_{(y,z)\in H} \sup_{\substack{j_1,j_2 \\ 0\le j_1+j_2\le r}} \{\|D_z^{j_1} D_y^{j_2} \Phi(y,z)\| + \|D_z^{j_1} D_y^{j_2} \Psi(y,z)\|\}$$

for any pair of functions Φ and Ψ defined on H and C^r jointly in y and z. Then $\lambda(\overline{Y},\overline{Z})$ is small if ε is small.

For $\varepsilon_1 > \varepsilon > 0$ the system

$$\begin{aligned} y' &= Cy + \overline{Y}(y,z) \\ z' &= Az + \overline{Z}(y,z) \end{aligned} \tag{***}$$

has the same properties local to $(0,0)$ as does the system (*), with $\varepsilon y = x_c$ and $\varepsilon z = x_s$. For $0 < \varepsilon < \varepsilon_1$ the system (***) is defined globally and generates a C^r semiflow $\phi_t(y_0,z_0) = (y_t,z_t)$.

Define

$$\overline{Y}_{t,\varepsilon}(y_0,z_0) = y_t - e^{Ct}y_0$$

and

$$\overline{Z}_{t,\varepsilon}(y_0,z_0) = z_t - e^{At}z_0$$

<u>Lemma</u>. <u>The following statements hold uniformly over</u> $0 \le t \le 1$:

a) $\overline{Y}_{t,\varepsilon} = 0$, $\overline{Z}_{t,\varepsilon} = 0$ <u>if</u> $\|y_0\| > \frac{1}{2}$ <u>for small</u> ε.

b) $\lambda(\overline{Y}_{t,\varepsilon},\overline{Z}_{t,\varepsilon}) \to 0$ <u>as</u> $\varepsilon \to 0$.

The time one map is $\phi_1(y_0,z_0) = (y_1,z_1)$. The existence of an invariant manifold for this map is guaranteed by the following:

<u>Lemma</u>. (Lemma (2.4) in Marsden-McCracken [81, p. 32]. <u>If</u> δ <u>is close enough to</u> 0, <u>then for sufficiently small</u> ε, <u>there exists an invariant manifold of</u> ϕ_1 <u>that is defined by a</u> C^{r-1}

function $z = g_\varepsilon(y)$ *with* $g_\varepsilon(0) = 0$, $Dg_\varepsilon(0) = 0$, *and* $\|Dg_\varepsilon(y)\| < 1$. *Furthermore,* $\|z_n - g_\varepsilon(y_n)\| \to 0$ *as* $n \to \infty$, *where* $(y_n, z_n) = \phi_1^n(y_0, z_0) = \phi_n(y_0, z_0)$.

Fix any such small ε and δ , say $\varepsilon = \varepsilon_0$, $\delta = \delta_0$ with $c - a > 4\lambda$, where $a = \|e^A\|$, $1/c = \|e^{-C}\|$. One can show, following Hartman's proof of existence of invariant manifolds [43, Chapter IX], that

(α) if $z_0 = g_{\varepsilon_0}(y_0)$, then $\|y_m\| \geq (c - 2\lambda)^m \|y_0\|$ for $m = 1,2,\ldots$

(β) $\|v_m\| \leq (a + 2\lambda)^m \|v_0\|$ for $m = 1,2,\ldots$, where $v_t = z_t - g_{\varepsilon_0}(y_t)$.

Furthermore, one can show that

(γ) For $c > 2\lambda$ the restriction of ϕ_1 on the invariant manifold $z = g_{\varepsilon_0}(y)$ is a diffeomorphism (onto).

This last result follows from:

Proposition. Let $\varphi: R^n \to R^n$ *be a* C^1 *map of the form*

$$\varphi(y) = By + G(y) ,$$

where B *is a nonsingular constant matrix. If* $\|B^{-1}\| \sup_y \|DG(y)\| < 1$, *then* φ *is a diffeomorphism onto.*

To establish (γ) one applies this Proposition to the time one map

$$\phi_1(y) = e^C y + \bar{Y}_{1,\varepsilon_0}(y, g_{\varepsilon_0}(y)) ,$$

which has the form

$$\phi_1(y) = By + G(y) .$$

To outline a proof of part (a) of the Center Manifold Theorem, it suffices to show the manifold $z = g_{\epsilon_0}(y)$ is also invariant under the maps ϕ_{t_0} for t_0 between 0 and 1. Let $z_0 = g_{\epsilon_0}(y_0)$. By ($\gamma$), one can find (y_{-n}, z_{-n}) for $n = 1,2,\ldots$ such that $z_{-n} = g_{\epsilon_0}(y_{-n})$ and $\phi_n(y_{-n}, z_{-n}) = (y_0, z_0)$. By $\|Dg(y)\| < 1$, $g(0) = 0$ and (β), one has

$$\begin{aligned} \|z_{-n+t_0}\| + \|y_{-n+t_0}\| &\geq \|z_{-n+t_0}\| + \|g(y_{-n+t_0})\| \\ &\geq \|v_{-n+t_0}\| \geq (a + 2\lambda)^{-n}\|v_{t_0}\| . \end{aligned} \tag{i}$$

Since $\|y_{-n+t_0} - e^{Ct_0} y_{-n}\| \leq \lambda(\|y_{-n}\| + \|z_{-n}\|)$

and

$$\|z_{-n+t_0} - e^{At_0} z_{-n}\| \leq \lambda(\|y_{-n}\| + \|z_{-n}\|) ,$$

one gets

$$\begin{aligned} \|y_{-n+t_0}\| + \|z_{-n+t_0}\| &\leq 2[\|y_{-n}\|e^{-\alpha} + \lambda(\|y_{-n}\| + \|y_{-n}\|)] \\ &\leq 2\|y_{-n}\| [e^{-\alpha} + 2\lambda] . \end{aligned} \tag{ii}$$

By (i), (ii), and (α),

$$2(e^{-\alpha} + 2\lambda)(c - 2\lambda)^{-n} \|y_0\| \geq (a + 2\lambda)^{-n} \|v_{t_0}\| ,$$

or

$$2(e^{-\alpha} + 2\lambda)(a + 2\lambda)^{n}(c - 2\lambda)^{-n} \|y_0\| \geq \|v_{t_0}\| .$$

Let $n \to \infty$. Then since $c - a > 4\lambda$, one obtains $v_{t_0} = 0$; or $z = g_{\epsilon_0}(y)$ is invariant under the map ϕ_{t_0} _for any_ t_0 _between_ 0 _and_ 1.

Let

$$\theta = \sup_{(y,z)\in H} \{\|D_y\overline{Y}(y,z)\|, \|D_y\overline{Z}(y,z)\|, \|D_z\overline{Y}(y,z)\|, \|D_z\overline{Z}(y,z)\|\} .$$

Clearly $\theta \to 0$ as $\varepsilon \to 0$. Thus, we may assume that $\alpha + 2\theta < 0$.

To prove part (b), writing $v(t)$ for v_t for convenience, we will establish

$$\|v(t)\| \le \|v(0)\|e^{(\alpha+2\theta)t} \quad \text{for} \quad t \ge 0 .$$

The set of initial points of C^1 solutions of the system (***) is dense in H (see [1, Chapter 5; 48, Chapter 3]). Thus, one can assume $(y(t), z(t))$ is C^1 for $t \ge 0$. By (***) and the definition of v (see (β)) the operator equation for v is $v' = Av + V(y,v)$, where $V(y,0) = 0$ and

$$V(y,v) = Ag(y) + \overline{Z}(y,v+g(y)) - (Dg)(Cy + \overline{Y}(y,v+g(y))).$$

Now,

$$\|V(y,v)\| \le \Big(\sup_{(y,v)\in H} \|D_vV(y,v)\|\Big) \|v - 0\| ,$$

$$\|D_vV(y,v)\| = \|D_z\overline{Z}(y,v+g(y)) - (Dg)D_z\overline{Y}(y,v+g(y))\| \le 2\theta ,$$

in which the last inequality follows from $\|Dg\| < 1$. Thus $\|V\| \le 2\theta\|v\|$. Let $h(t) = e^{-\alpha t}\|v(t)\|$. The equation

$$v(t) = e^{At}v(0) + \int_0^t e^{A(t-s)} V(y(s),v(s))ds$$

implies that

$$h(t) \le h(0) + \int_0^t (2\theta) h(s)ds .$$

By Gronwall's inequality [43, p. 24], one obtains $h(t) \le h(0)\exp(2\theta)t$ or $\|v(t)\| \le e^{(d+2\theta)t}\|v(0)\|$ for $t \ge 0$. Thus, the outline of our proof is completed.

When one considers a bifurcation problem $x' = f(x,\mu)$ near $\mu = 0$ such as the Hopf bifurcation, one can apply the center manifold theorem to the suspended system

$$\begin{cases} x' = f(x,\mu) \\ \mu' = 0 \end{cases} \qquad (\dagger)$$

at the origin $(0,0)$. The center manifold can be represented as $x_s = g(x_c,\mu) = g_\mu(x_c)$ and the restriction of the suspended system to its center manifold is:

$$\begin{cases} x_c' = f^*(x_c,\mu) \\ \mu' = 0 \end{cases}, \quad \text{where } x = (x_c,x_s) . \qquad (\dagger\dagger)$$

If $(x_c(\cdot),\mu)$ is a solution of $(\dagger\dagger)$, then $(x_c(\cdot),g_\mu(x_c(\cdot),\mu)$ is a solution of $(\dagger)$. By the local invariance property of the center manifold, a bifurcating family of periodic solutions for $(\dagger)$ necessarily belongs to the center manifold and therefore corresponds to bifurcating periodic solutions for $(\dagger\dagger)$.

<u>Proposition</u>. <u>For</u> μ <u>sufficiently small, bifurcating periodic solutions of</u> $x' = f(x,\mu)$ <u>near the origin are asymptotically stable if the corresponding solutions of</u> $x_c' = f^*(x_c,\mu)$ <u>are asymptotically stable solutions for</u> $x_c' = f^*(x_c,\mu)$.

<u>Outline of Proof</u>.

Set $w_\mu = x_s - g_\mu(x_c)$. By local attractivity of the center manifold (as established in the outline of proof of the Center Manifold Theorem), one can find a neighborhood V of 0 in H, $\mu_0 > 0$, and κ $(0 < \kappa < 1)$, such that $\|w_\mu(t)\| \le \kappa^t \|w_\mu(0)\|$ for $t \ge 0$, provided $w_\mu([0,t]) \subset V$ and $|\mu| \le \mu_0$. Consider a sufficiently small μ with the properties:

1) $|\mu| \le \mu_0$,
2) the bifurcated periodic solution γ_μ of $(\dagger\dagger)$ lies in V,
3) the corresponding periodic solution of the equation $x_c' = f^*(x_c,\mu)$ has a Lyapunov function $\phi_\mu(x_c)$ [82].

While the Lyapunov function ϕ_μ is defined with respect to the system $(\dagger\dagger)$, we now consider $\phi_\mu(t) = \phi_\mu(x_c(t))$ where x_c

is the finite dimensional component of a solution of the original system (†). Next, for $\delta > 0$ and $\eta > 0$ (δ and η small) we define

$$B_{\delta,\eta} = \{(x_c,x_s) \mid \|w_\mu\| \le \eta \text{ and } \phi_\mu \le \delta\} .$$

Fix $\delta = \delta_1 > 0$ so that $B_{\delta_1,0} \subset V$ and $\phi'_\mu \le 0$ on $B_{\delta_1,0}$ ($\phi'_\mu = 0$ only on the periodic orbit γ_μ). We next prove that if η is chosen to be small enough, then $B_{\delta_1,\eta}$ is positively invariand with respect to (†). First, on that portion of $\partial B_{\delta_1,\eta}$ where $\|w_\mu\| = \eta$, each trajectory of (†) enters $B_{\delta_1,\eta}$ since M is locally attracting, provided η is small enough. Second, let S be the remaining portion of $\partial B_{\delta_1,\eta}$, where $\|w_\mu\| < \eta$ but $\phi_\mu = \delta_1$. Then $\phi'_\mu(t) < 0$ on S. To see this, recall that in V, $x'_c = Cx_c + Y(x_c,x_s)$. By direct computation,

$$\begin{aligned}\phi'_\mu(t) &= \operatorname{grad}\phi_\mu \cdot x'_c \\ &= \operatorname{grad}\phi_\mu \cdot [Cx_c + Y(x_c,x_s)]\end{aligned}$$

so that ϕ'_μ always exists and is continuous in $x = (x_c,x_s)$. By adding and subtracting $Y(x_c,g_\mu(x_c))$ to $\phi'_\mu(t)$, we obtain

$$\phi'_\mu(t) = \phi'_\mu(t)|_{M\cap S} + \operatorname{grad}\phi_\mu \cdot [Y(x_c,x_s) - Y(x_c,g_\mu(x_c))] .$$

But on $M\cap S$, ϕ'_μ is strictly negative, independent of η, say $\phi'_\mu(t)|_{M\cap S} \le -c(\delta_1) < 0$, where $c(\delta)$ decreases monotonically to 0 as δ does, since on S, $\|x_c - \gamma_\mu\| \ge k_1 > 0$ for some $k_1 > 0$. Furthermore, for some $K > 0$,

$$\|Y(x_c,x_s) - Y(x_c,g_\mu(x_c))\| \le K\|x_s - g_\mu(x_c)\| \le K\eta .$$

Now choose $\eta = \eta(\delta) = c(\delta)/(2K)$. Then $\phi'_\mu(t) < 0$ on S if $\eta = \eta(\delta_1)$. Hence, $B_{\delta_1,\eta(\delta_1)}$ is positively invariant for (†).

Lastly, suppose that there exists a trajectory $x(t)$ of (†) that enters $B_{\delta_1,\eta(\delta_1)}$ but that does not enter each $B_{\delta,\eta(\delta)}$ for $\delta \in (0,\delta_1)$. But if $x(t)$ does not enter $B_{\delta_2,\eta(\delta_2)}$, say, with

$0 < \delta_2 < \delta_1$, then (1) x(t) cannot remain in that portion of

$$D = B_{\delta_1,\eta(\delta_1)} \setminus B_{\delta_2,\eta(\delta_2)} ,$$

where $\|w_\mu\| > \delta_2$, for more than a finite time since M is exponentially attracting, and (2) x(t) cannot remain in the rest of D for more than a finite time either, because this would imply that $\phi_\mu(t) \to -\infty$ as $t \to \infty$, which is impossible. Since $B_{\delta_1,\eta(\delta_1)}$ is positively invariant we have a contradiction; hence each solution of (†) with initial point on $\partial B_{\delta_1,\eta(\delta_1)}$ enters every $B_{\delta,\eta(\delta)}$ for $\delta \in (0,\delta_1)$. Thus the periodic solution of (†) corresponding to γ_μ is asymptotically, orbitally stable fo μ sufficiently small. This completes the proof of the above proposition.

Remark. The reader should note that it is in terms of the norm o: the Hilbert space that stability is defined.

We thank our colleague Matthew Bottkol for criticisms and suggestions which improved this appendix.

Appendix B. Summary of Results on Continuous Semigroups

In this Appendix, we define contraction, analytic, and compact semigroups and give characterizations of their infinitesimal generators. The main references are Hille-Phillips [49], Balakrishnan [8], Friedman [34], and Mizohata [85]. Perturbation results about these semigroups are also presented. The main reference is Kato [69].

Let X be a complex Banach space with norm $| \ |$. For any bounded linear transformation $B: X \to X$, $\|B\| = \sup_{|x|\leq 1} |Bx|$ denotes the operator norm of B. Given any linear transformation $A: D_A \to X$, $\rho(A)$ stands for the resolvent set, and $R(\lambda,A) = (\lambda I - A)^{-1}$ the resolvent of A for any λ in $\rho(A)$. Throughout this Appendix, a linear C^0 semigroup will be simply called a C^0 semigroup.

Let T_t $(0 \leq t < \infty)$ be a C^0 semigroup on a Banach space X. Let D_A be the subspace of all elements x such that $\frac{T_h(x) - x}{h}$ converges as $h \to 0+$. The infinitesimal generator $A: D_A \to X$ of the semigroup T_t is the linear operator defined by $Ax = \lim_{h\to 0+} \frac{T_h(x) - x}{h}$. Given a linear operator A, we say that it generates a C^0 semigroup if A coincides with the infinitesimal generator of some C^0 semigroup.

<u>Proposition</u> 1. <u>Let</u> T_t <u>be</u> <u>a</u> C^0 <u>semigroup</u> <u>with</u> <u>infinitesimal</u> <u>generator</u> A. <u>Then</u>,

1) A <u>is</u> <u>a</u> <u>closed</u> <u>linear</u> <u>operator</u> <u>with</u> <u>dense</u> <u>domain</u>.

2) <u>For</u> <u>any</u> $x \in D_A$, $T_t(x) \in D_A$ <u>for</u> <u>all</u> $t > 0$ <u>and</u> $\frac{d}{dt} T_t(x) = AT_t(x) = T_tA(x)$.

<u>Theorem</u> 1. (Hille-Yosida-Phillips). <u>A</u> <u>necessary</u> <u>and</u>

sufficient condition that a closed linear operator A with dense domain D_A be the infinitesimal generator of a C^0 semigroup is that there exist real numbers M, ω $(M > 0)$ such that for every λ in $\rho(A)$ with $\mathrm{Re}\lambda > \omega$, the inequalities

$$\|R(\lambda,A)^n\| \leq M/(\mathrm{Re}\lambda - \omega)^n \qquad (*)$$

hold for $n = 1,2,\ldots$.

It follows from the proof that $\|T_t\| \leq Me^{\omega t}$ for all $t \geq 0$ with the same M, ω as in equation (*).

Theorem 2. Let S_t, T_t be two C^0 semigroups having the same infinitesimal generator, then $S_t = T_t$ for all $t \geq 0$.

A semigroup T_t is called a contraction semigroup if $\|T_t\| \leq 1$ for all $t \geq 0$. A necessary and sufficient condition for a closed operator A with dense domain to generate a C^0 contraction semigroup is that $\|R(\lambda,A)\| \leq \frac{1}{\lambda}$ for all $\lambda > 0$ [8; p. 171]. When X happens to be a Hilbert space with inner product $\langle , \rangle$, one has simple criteria for generators of contraction semigroups. A closed linear operator A with dense domain D_A is said to be dissipative if $\langle Ax,x\rangle + \langle x,Ax\rangle \leq 0$ for all x in D_A.

Theorem 3 [8; pp. 174-175]. (a) Let T_t be a contraction semigroup over a Hilbert space H. Then, its infinitesimal generator A is dissipative.

(b) Suppose A is dissipative and the range $I - A$ is the whose space. Then, A generates a contraction semigroup.

(c) Suppose both A and its adjoint A^* are dissipative. Then, A generates a contraction semigroup.

To deal with t-smoothness of C^0 semigroups the concept of an analytic semigroup is relevant. Given θ in $(0, \frac{\pi}{2})$ set $\Delta_\theta = \{\xi \in C \mid \xi \neq 0$ and $|\text{argument of } \xi| < \theta\}$. A semigroup T_t in X is said to be analytic if it can be extended to a family

T_ξ of bounded linear transformations with ξ lying in Δ_θ for some θ in $(0, \frac{\pi}{2})$, such that

(i) $T_{\xi_1+\xi_2} = T_{\xi_1} \cdot T_{\xi_2}$

(ii) $T_\xi(x)$ is analytic in the sector Δ_θ for each x in X

(iii) $|T_\xi(x) - x| \to 0$ as $|\xi| \to 0$ in any closed subsector of Δ_θ, for any x in X.

Proposition 2. *Suppose T_t is an analytic semigroup with infinitesimal generator A. Then, to each $t > 0$,*

) $T_t(X) \subset$ domain of A^n, *for* $n = 1,2,\ldots$

) $\frac{d^n T_t}{dt^n} = A^n T_t$ *in the operator topology on bounded transformations for* $n = 1,2,\ldots$ *In particular,* $t \to \|T_t\|$, $t \to \|AT_t\|$ *are continuous in* t *on* $(0,\infty)$.

Let $\phi \in (0, \frac{\pi}{2})$ and $a > 0$ be given.

Theorem 4. *Suppose that A is a closed operator with dense domain such that*

) *The resolvent set of A contains the sector*

$$S_{\phi,a} = \{\lambda \in \mathbb{C} \mid \lambda \neq a, |\arg(\lambda - a)| < \frac{\pi}{2} + \phi\}$$

) $\|R(\lambda,A)\| \leq M/|\lambda|$ *if* $\lambda \in S_{\phi,a}$

Then, A generates an analytic semigroup which can be extended to the sector Δ_ϕ.

A C^0 semigroup of bounded linear operators T_t is said to be *compact* if T_t is compact for all $t > 0$.

Theorem 5. *A C^0 semigroup T_t is compact is and only if*

) $t \to \|T_t\|$ *is continuous on* $(0,\infty)$

) $R(\lambda,A)$ *is compact for some* λ *in* $\rho(A)$.

(Hence, $R(\lambda,A)$ is compact for any λ in $\rho(A)$).

It follows from this theorem that any compact semigroup has the following properties:

1) The infinitesimal generator A has a pure point spectrum consisting at most a countable sequence of points $\{\lambda_k\}$ with corresponding eigenvectors $\{\phi_k\}$, and $\{\lambda_k\}$ cannot have an accumulation point in the finite part of the plane.

2) $T_t\phi_k = e^{\lambda_k t}\phi_k$

3) The spectrum of T_t = the closure of $\{e^{\lambda_k t} \mid k = 1,2,\ldots\}$ for $t > 0$; (i.e., $\sigma(T_t) = \overline{\exp(t\sigma(A))}$).

In conclusion, we quote some perturbation results of semigroups from Kato [69].

Theorem 6. *Suppose* A *generates a* C^0 *semigroup on a Banach space* X, *and* B *is a bounded transformation on* X. *Then,*

1) $A + B$ *also generates a* C^0 *semigroup on* X.

2) *If* A *generates an analytic semigroup, so does* $A + B$.

3) *If* A *generates a compact semigroup, so does* $A + B$.

Here, of course, $A + B$ denotes the linear transformation from the domain of A into X which is defined by $(A + B)(x) = Ax + Bx$.

A family L_μ of closed operators on X, defined for μ in a domain D_0 of the complex plane, is said to be *holomorphic of type* (A) if (1) the domain $D(L_\mu) = D$ is independent of μ and (2) $L_\mu x$ is holomorphic in μ for every x in D.

Theorem 7. (Kato [69, Chapter IX, Th. 2,6]). *Let* L_μ *be holomorphic family of type* (A) *defined near* $\mu = 0$. *If* L_0 *is the generator of an analytic semigroup, the same is true for* L_μ *with* $|\mu|$ *sufficiently small. In this case,* $U(t,\mu) = e^{tL_\mu}$ *is holomorphic in* μ *and* t *when* t *is in some open sector ontaining* $t > 0$. *Moreover*, *all* $\frac{\partial^n U}{\partial \mu^n}$ *are strongly continuous*

n t *up to* $t = 0$.

Appendic C. A Regularity Theorem

Let Ω be a bounded domain with smooth boundary in $\mathbb{R}^n$. Recall that $H^\ell(\Omega,\mathbb{R})$ denotes the real Sobolev space introduced at the beginning of Section 2 in Chapter 5.

<u>Proposition 1</u>. <u>Assume that</u> $f\colon \Omega\times\mathbb{R}^m \to \mathbb{R}$ <u>is a smooth function, and suppose</u> $\ell > \frac{n}{2}$. <u>Then, the map</u> $F\colon s \to f(\cdot,s(\cdot))$ <u>from</u> $\oplus^m H^\ell(\Omega,\mathbb{R}) \to H^\ell(\Omega,\mathbb{R})$ <u>is well-defined and smooth</u>.

To verify this well-known proposition [1], the following lemma from Palais [90] at page 31 is needed:

<u>Lemma</u>. <u>Assume that</u> $g\colon \Omega\times\mathbb{R}^m \to \mathbb{R}$ <u>is a smooth function, and suppose</u> $\ell > \frac{n}{2}$. <u>Then the map</u> $s \to g(\cdot,s(\cdot))$ <u>from</u> $\oplus^m H^\ell(\Omega,\mathbb{R}) \to H^\ell(\Omega,\mathbb{R})$ <u>is well-defined and continuous</u>.

Proof of Proposition 1. By the Taylor Theorem, for any r,

$$f(x,s+h) = \sum_{|\alpha|=0}^{r} \frac{\phi_\alpha(x,s)}{\alpha!} h^\alpha + \sum_{|\alpha|=r} R_\alpha(x,s,h)h^\alpha ,$$

with ϕ_α, R_α smooth in their arguments and $R_\alpha(x,s,0) = 0$. From the above Lemma, the maps

$$h \to h^\alpha(\cdot),\ s \to \phi_\alpha(\cdot,s(\cdot)),\quad s,h \to R_\alpha(\cdot,s(\cdot),h(\cdot))$$

are continuous in the norms $\|\ \|_\ell$ induced by $\|\ \|_\ell^\Omega$. Thus, the multilinear map, defined by $\frac{\phi_k(s)}{k!}(h) = \sum_{|\alpha|=k} \frac{\phi_\alpha(\cdot,s(\cdot))}{\alpha!} h^\alpha(\cdot)$ is bounded and $s \to \Phi_k(s)$ is continuous from $H^\ell(\Omega,\mathbb{R})$ into $L_s^k(H^\ell(\Omega,\mathbb{R}),H^\ell(\Omega,\mathbb{R}))$. Since,

$R_\alpha(\cdot,s(\cdot),h(\cdot)) \to R_\alpha(\cdot,s(\cdot),0) = 0$ as $(s,h) \to (s,0)$. One gets $\frac{\|R(s,h)\|_\ell}{\|h\|_\ell^r} \to 0$ as $(s,h) \to (s,0)$, where $R(s,h) = \sum_\alpha R_\alpha(\cdot,s(\cdot),h(\cdot))h^\alpha(\cdot)$. Clearly,

$$F(s+h) = \sum_{k=0}^{r} \frac{\Phi_k(s)}{k!} h^k + R(s,h)h^r .$$

By a converse of Taylor's Theorem in Banach space (see Theorem 2.1 at page 6 in Abraham and Robbin [1]). One obtains that $F: \oplus^m H^\ell(\Omega,\mathbb{R}) \to H^\ell(\Omega,\mathbb{R})$ is of class C^r. Now r can be any positive integer, F is, therefore, smooth.

Appendix D. Truncation Error, Roundoff Error and Numerical Differencing

Our object in this Appendix is to explain the various choices of increments made in Chapter 3. The concepts are standard, see for example [27, 112].

Suppose that a number g_0 is defined by $g_0 = \lim_{h\to 0} g(h)$, where $g(0)$ may or may not be defined. The <u>truncation error</u> $t(h)$ in approximating g_0 by $g(h)$ is then

$$t(h) = g_0 - g(h) .$$

(The expression "truncation error" is more natural in the context of series, where the partial sum of N terms is the "truncated" series. In the present context, the limit $h \to 0$ is "truncated" by stopping at a nonzero h.)

Now suppose that $g(h)$, $h \neq 0$, is evaluated numerically rather than exactly, and let $G(h)$ denote the machine result. The <u>roundoff error</u> in evaluating $g(h)$ is

$$r(h) = g(h) - G(h) .$$

(A machine only carries numbers to a certain finite precision, and so during computations must approximate those numbers which cannot be represented exactly. The cumulative effect of these errors is termed "roundoff error" because rounding is a common technique, although not the only one, for approximating numbers by machine representable numbers.) Note that if $G(h)$ is not identically constant, it has jump discontinuities since its values are machine representable numbers.

The <u>total error</u> in approximating g_0 by $G(h)$ is then the sum of the truncation and roundoff errors:

$$e(h) = g_0 - G(h) = t(h) + r(h) .$$

The question arises: how best to choose the value of h to minimize the total error? Often there is not enough information available to answer this question, and so one looks instead for the value of h that minimizes the sum

$$T(h) + R(h) ,$$

where $T(h)$ and $R(h)$ are bounds for $t(h)$ and $r(h)$, respectively.

Suppose

$$T(h) = C_1 h^p , \quad R(h) = C_2 u / h^q ,$$

where $p > 0$, $q > 0$, u is the machine precision (the smallest number such that the machine distinguishes between $(1.0 + u)$ nd (1.0)), and C_1 and C_2 are positive constants independent of h and u. The sum $T(h) + R(h)$ is minimized for

$$h = \left(\frac{q}{p} \frac{C_2}{C_1} u \right)^{1/(p+q)} .$$

For the one-sided difference approximation

$$g(h) = (f(x_0 + h) - f(x_0))/h \approx g_0 = f'(x_0) ,$$

$= 1$ and $q = 1$. However, the ratio C_2/C_1 must be treated s an unknown. For C_1 involves $f''(x_0)$, and since the result f the computation is an approximation to $f'(x_0)$, it would be nreasonable to assume that $f''(x_0)$ is known. Moreover, C_2 nvolves the actual mechanism of accumulation of roundoff error, which is often inadequately known. In the absence of this dditional information, the choice

$$h = x_{ref} \, u^{1/2}$$

an be made, where x_{ref} is a scale for the variable x. For his choice, the total error is $O(u^{1/2})$. The "O" symbol here refers to the limit as one performs the same computation using floating point arithmetic of successively greater precision. Although a somewhat unrealistic limit, the power $u^{1/2}$ indicates that under "normal" circumstances, one-sided numerical differencing can produce a value for $f'(x_0)$ with roughly 1/2 the number of significant figures carried by the machine.

For the two-sided difference approximation

$$g(h) = (f(x_0 + h) - f(x_0 - h))/2h \approx g_0 = f'(x_0) ,$$

$p = 2$ and $q = 1$ and the ratio C_1/C_2 must again be treated as unknown since C_1 involves $f'''(x_0)$. An increment $h = x_{ref} u^{1/3}$ is then appropriate, for which choice the total error in the approximation to the derivative is $O(u^{2/3})$.

For the three point approximation

$$g(h) = (f(x_0 + h) + f(x_0 - h) - 2f(x_0))/h^2 \approx g_0 = f''(x_0) ,$$

the exponents are $p = q = 2$, and the choice $h = x_{ref}u^{1/4}$ is appropriate, for which the total error is $O(u^{1/2})$.

Let

$$\Delta_5(h)f(x_0,y_0) = [f(x_0 + h, y_0) + f(x_0 - h, y_0) + f(x_0,y_0 + h) + f(x_0,y_0 - h) - 4f(x_0,y_0)]/h^2 .$$

This is the "5-point Laplacian". For the approximation

$$g(h) = \Delta_5(h)f(x_0,y_0) \approx g_0 = \Delta f(x_0,y_0) ,$$

the exponents are again $p = q = 2$, and the choice $h = s_{ref}u^{1/4}$ is appropriate, where s_{ref} is a common scale for the variables x and y .

Let

$$\Delta_9(h)f(x_0,y_0) = \frac{1}{3}[4\,\Delta_5(2h)f(x_0,y_0) - \Delta_5(h)f(x_0,y_0)]$$

This is the "9-point Laplacian", and may be thought of as having been constructed from the 5-point Laplacian by the process of Richardson extrapolation [27], which has the effect of increasing p from 2 to 4. The exponents are $p = 4$ and $q = 2$ so the choice $h = s_{ref}u^{1/6}$ is appropriate, for which the total error is $O(u^{2/3})$.

The first step in the secant iteration for location for location of ν_c (Chapter 3, Section 2) may be thought of as a Newton step, in which a one-sided difference approximation is used for the derivative $\alpha'(\nu_1)$ and the size of the increment employed is an informed guess of the value required to minimize the total error in the approximation. Note that the computed value of ν_c will be (essentially) independent of the scheme used to start the secant iteration.

A decision was made that the error due to numerical differencing in all of $c_1(0)$, $\alpha'(0)$, $\omega'(0)$, μ_2, τ_2 and β_2, should be $O(u^{2/3})$. Therefore two-sided (rather than one-sided) differencing is used in computing f_{20} and f_{11} (Chapter 3, Section 3.3), the 9-point (rather than 5-point) Laplacian is used in computing G_{21} (Chapter 3, Section 3.6), and two-sided (rather than one-sided) differencing is used in computing $\lambda_1'(\nu_c)$ (Chapter 3, Section 4). Note that if the decision were to accept an $O(u^{1/2})$ error, the number of Jacobian evaluations in this phase of the computation could be reduced, roughly by a factor of 2.

Appendix E. The Code BIFOR2

1. Introduction.

The code BIFOR2 is the product of several years of evolution A brief history of the code will perhaps help others avoid some of the pitfalls we encountered.

Our first idea was to use a language for symbolic manipulation in order to perform the tedious algebraic manipulations which made hand calculations impossible. This approach was employed both in the derivation of bifurcation formulae [46] and in the application [44]. With symbolic manipulation, however, there were numerous technical problems. The number of potentially distinct analytic expressions involved in the term g_{21} is at least $N^2(N^2 + 3N + 2)/6$, the number of potentially distinct third order partial derivatives $\partial^3 f^i/\partial x_j \partial x_k \partial x_l$, $i,j,k,l = 1,\ldots,N$. The program employed in [44] was therefore limited to relatively low order systems. Also, because of the hybrid symbolic/numeric nature of the program, it was tied to the particular machine used (a CDC 6400) and to the peculiar combination of languages employed (SYMBAL, SNOBAL, FORTRAN).

A new version, was therefore written, entirely in FORTRAN [45]. The symbolic manipulation phase was eliminated, the partial derivatives being calculated instead by numerical differentiation. Once symbolic manipulation had been eliminated, a striking economy became apparent. When the system is transformed into real canonical form, only $O(N)$ distinct second partial derivatives and $O(1)$ distinct third partial derivatives are needed to evaluate the bifurcation formulae. The new version therefore performed the numerical differentiation in the canonical coordinate system rather than in the original coordinates. Program storage requirements were reduced to $O(N^2)$, and the program became applicable to higher order systems. Because there is a certain loss of accuracy associated with numerical differentiation, procedures for estimation of this error were built into the code. Also, preliminary checking of the user-supplied subroutine for consistency between the Jacobian matrix and the function values was incorporated because such checks had proven useful in the study [44]. BIFOR1 followed the "recipe" given in Chapter 2, and was used to verify the analytical results presented there. Every disagreement between BIFOR1 and the hand calculations was traced to mistake(s) in the latter.

Although able to analyze systems much more difficult than could be treated by hand or with the aid of symbolic manipulation, the code BIFOR1 was soon pressed to the limit of its capabilities in the panel flutter problem, and the need for increased efficiency became apparent. A new version, BIFOR2, was therefore written.

In BIFOR2, the explicit construction of the matrices P and P^{-1} of the transformation to real canonical coordinates was eliminated in favor of the technique employing right and left eigenvectors as described in Chapter 3. Also, the numerical differentiation procedure was reorganized into two distinct stages: the reorganization reduced the number of Jacobian evaluations required during the numerical differencing from $O(N)$ to $O(1)$. Different techniques for location of the critical value of ν were introduced. The two techniques (MTH = 1 and 2) retained in the current version are not as fast as a third technique which we developed, based upon solving the $N + 2$ (real) dimensional system

$$f(x; \nu) = 0$$

$$\det \left(\frac{\partial f}{\partial x}(x; \nu) - i\omega I\right) = 0$$

simultaneously for $(x(\nu_c), \nu_c, \omega(\nu_c))$. This third technique employed a quasi-Newton method in which the first N rows of the Jacobian matrix of the N + 2 dimensional system were evaluated explicitly but in which the last 2 rows were approximated by means of a 2 by N + 2 dimensional matrix B which was rank 1 updated at each iterate. The difficulty with this third technique was the initialization of the matrix B. If done carefully (all columns of B evaluated by numerical differencing), the overall scheme became too expensive. If done less carefully (columns 1, N + 1 and N + 2 evaluated by differencing, the remaining columns set to 0), the algorithm would locate ν_c efficiently for all the problems considered in Chapter 3, but could be fooled (made extremely slow) simply by changing the numbering of the individual differential equations within the system. The third technique has not been included in the present code BIFOR2 because we do not consider that the gain in speed (up to a factor of 2 on the problems in Chapter 3) offsets the potential loss of reliability.

Final improvements in BIFOR2 were the reduction in array storage requirements to the present $2N^2 + 15N$ locations, and the replacement of IMSL routines with the equivalent EISPACK [100] and LINPACK [29] routines for the algebraic eigenvalue problem and for the solution of linear systems. The use of EISPACK and LINPACK allows us to publish and distribute the complete code BIFOR2 without copyright concerns.

2. Partial listing of BIFOR2

A partial listing of BIFOR2 which describes the parameters of the subroutine is given below. For the full listing, see the microfiche.

```
 SUBROUTINE BIFOR2(FNAME,XS,N,ANU,U,MTH,JJOB,IPRINT,
1     PAR,V1,ERR,W,IER)

FUNCTION                 - ANALYSIS OF HOPF BIFURCATION IN AN
                           ORDINARY DIFFERENTIAL SYSTEM
                              DX/DT = F(X;ANU) .
                           THIS SUBROUTINE FIRST LOCATES A
                           CRITICAL VALUE ANUC OF THE BIFURCATION
                           PARAMETER, THEN COMPUTES AMU2, TAU2,
                           AND BETA2.  THE FAMILY OF PERIODIC
                           SOLUTIONS IS GIVEN BY
                           XS(ANUC)+EPS*RE(EXP(2*PI*T/PERIOD)*V1)
                                         + O(EPS**2)
                           THE PERIOD IS
                              (2*PI/OMEGA(0))*(1 + TAU2*EPS**2)
                                      + O(EPS**4),
                           AND THE CHARACTERIC EXPONENT
                           WHICH DETERMINES THE STABILITY OF
                           THE PERIODIC SOLUTIONS IS
                              BETA2*EPS**2 + O(EPS**4).
                           THE RELATION BETWEEN EPS AND ANU IS
                           ANU = ANUC + AMU2*EPS**2 + O(EPS**4).

AUTHOR OF BIFOR2         - B. HASSARD, DEPT. OF MATHEMATICS,
                           SUNY AT BUFFALO, BUFFALO N.Y. 14214
                           LAST REVISION APR. 2, 1980.

PARAMETERS     FNAME     - FNAME IS THE NAME OF THE USER-SUPPLIED
                              SUBROUTINE WHICH EVALUATES F(X;ANU)
                              AND THE JACOBIAN MATRIX. SEE BOX
                              BELOW FOR THE DESCRIPTION OF FNAME.

               XS        - XS IS AN N-VECTOR.
                           ON ENTRY, XS CONTAINS AN ESTIMATE OF
                              THE EQUILIBRIUM POINT FOR THE
                              CRITICAL VALUE OF ANU .
                           ON RETURN, XS IS THE EQUILIBRIUM POINT
                              FOR THE CRITICAL VALUE OF ANU.

               N         - N IS THE NUMBER OF FIRST ORDER
                              DIFFERENTIAL EQUATIONS, N .GE. 2.

               ANU       - ON ENTRY, ANU IS AN ESTIMATE OF THE
                              CRITICAL VALUE ANUC OF THE
                              BIFURCATION PARAMETER ANU .
                           ON RETURN, ANU IS THE CRITICAL VALUE.

               U         - U DETERMINES THE INCREMENTS USED IN
                              NUMERICAL DIFFERENCING. THE OPTIMAL
                              U DEPENDS UPON THE PROBLEM BUT WILL
```

```
C                         BE (APPROXIMATELY) THE SMALLEST
C                         POSITIVE NUMBER SUCH THAT THE
C                         MACHINE DISTINGUISHES BETWEEN
C                         (1.0+U) AND 1.0 .
C
C           MTH    - MTH DETERMINES THE METHOD USED TO
C                         LOCATE THE CRITICAL VALUE OF ANU.
C                      MTH = 1 USES THE SECANT METHOD TO ZERO
C                        RE EV1(ANU). FOR EACH VALUE OF ANU
C                         IN THE SECANT ITERATION, NEWTONS
C                         METHOD IS USED TO COMPUTE
C                         THE STATIONARY POINT XS(ANU),
C                         THEN THE QR ALGORITHM IS USED
C                         TO COMPUTE ALL THE EIGENVALUES
C                         OF THE JACOBIAN MATRIX.
C                      MTH = 2 IS SIMILAR, EXCEPT THAT
C                         AFTER THE FIRST EVALUATION OF EV1,
C                         INVERSE ITERATION IS USED TO
C                         COMPUTE JUST THE EIGENVALUE EV1.
C
C           JJOB   - IF JJOB = 0, EVALUATE BIFURCATION
C                         FORMULAE.
C                      IF JJOB = 1, IN ADDITION PERFORM
C                         ERROR ESTIMATION. (SEE ERR)
C
C           IPRINT - IPRINT DETERMINES THE AMOUNT OF
C                         OUTPUT TO LOGICAL UNIT 6.
C                      IPRINT = 0 FOR NO OUTPUT,
C                         (EXCEPT FOR ERROR MESSAGES ISSUED
C                         THROUGH SUBROUTINE UERTST)
C                      IPRINT = 1 FOR CRITICAL VALUES AND
C                         BIFURCATION PARAMETERS,
C                      IPRINT = 2 FOR EQUILIBRIUM POINT
C                         AND EIGENVALUE(S) OF JACOBIAN AT
C                         EACH ITERATE IN LOCATION OF ANUC.
C
C           PAR    - PAR IS A REAL 10-VECTOR, WHICH UPON
C                         RETURN CONTAINS THE RESULTS OF THE
C                         EVALUATION OF BIFURCATION FORMULAE.
C                      PAR(1) = AMU2 IS THE NUMBER THAT GIVES
C                         THE DIRECTION OF BIFURCATION.
C                      PAR(2) = TAU2 IS A COEFFICIENT IN THE
C                         EXPANSION OF THE PERIOD.
C                      PAR(3) = BETA2 IS THE LEADING
C                         COEFFICIENT IN THE EXPANSION OF A
C                         CHARACTERISTIC EXPONENT. THE
C                         PERIODIC SOLUTIONS ARE
C                         STABLE IF BETA2 .LT. 0,
C                         UNSTABLE IF BETA2 .GT. 0.
C                      PAR(4) = REAL(C1), WHERE C1 IS
```

```
C                         THE COEFFICIENT OF THE CUBIC TERM
C                         IN THE POINCARE NORMAL FORM,
C                         AT ANU = ANUC.
C                      PAR(5) = AIMAG(C1), C1 AS ABOVE.
C                      PAR(6) = DALPHA IS THE DERIVATIVE OF
C                         THE REAL PART OF EV1 WITH RESPECT
C                         TO ANU AT THE CRITICAL VALUE ANUC.
C                      PAR(7) = DOMEGA IS THE DERIVATIVE OF
C                         THE IMAGINARY PART OF EV1.
C                      PAR(8) = OMEGA, THE IMAGINARY PART OF
C                         EV1 AT THE CRITICAL VALUE.
C                      PAR(9) = ENORMX, THE EUCLIDEAN NORM OF
C                         THE STATIONARY POINT XS.
C                      PAR(10) = ENORMV, THE EUCLIDEAN NORM
C                         OF THE EIGENVECTOR V1.
C
C             V1     - V1 IS A COMPLEX N-VECTOR. UPON RETURN,
C                         V1 CONTAINS AN EIGENVECTOR OF THE
C                         JACOBIAN MATRIX AT THE STATIONARY
C                         POINT FOR THE CRITICAL VALUE .
C                         V1 CORRESPONDS TO EV1 = I*OMEGA.
C                         V1 IS NORMALIZED SO THAT ITS FIRST
C                         NON-VANISHING COMPONENT IS 1.0 .
C                         NOTE THAT IF V1 IS TO BE NORMALIZED
C                         INTO (SAY) E1 = V1/ENORMV AND THE
C                         APPROXIMATION TO THE PERIODIC
C                         SOLUTIONS WRITTEN IN TERMS OF E1,
C                         ALL OF AMU2, TAU2 AND BETA2
C                         MUST BE DIVIDED BY ENORMV**2 .
C
C             ERR    - ERR IS A REAL 7-VECTOR. UPON RETURN
C                         FOR JJOB = 1,  ERR(I) CONTAINS AN
C                         ESTIMATE OF THE ERROR DUE TO
C                         NUMERICAL DIFFERENCING IN PAR(I),
C                         WHERE I=1,...,7 . IF JJOB = 0,
C                         ERROR ESTIMATION IS NOT PERFORMED,
C                         AND ERR IS NOT USED.
C
C             W      - W IS A WORK AREA, AN ARRAY OF LENGTH
C                         AT LEAST 2*N**2 + 12*N .
C
C             IER    - ERROR PARAMETER
C                      IER = 0 INDICATES A NORMAL RETURN
C                      IER = 129 IF  N .LT. 2
C                      IER = 130 IF  U .LE. 0.0
C                      IER = 131 IF (MTH.LT.1).OR.(MTH.GT.2)
C                      IER = 132 IF (JJOB.LT.0)
C                                   .OR.(JJOB.GT.1)
C                      IER = 133 IF (IPRINT.LT.0)
C                                   .OR.(IPRINT.GT.2)
```

```
C                         IER = 134 - ERROR RETURN FROM CHECKJ,
C                            POSSIBLE INCONSISTENCY BETWEEN THE
C                            FUNCTION AND JACOBIAN AS EVALUATED
C                            BY FNAME.
C                         IER = 135 - ERROR RETURN FROM ANUCRT
C                            (LOCATION OF ANUC, MTH=1)
C                         IER = 136 - ERROR RETURN FROM ANUCRT
C                            (LOCATION OF ANUC, MTH=2)
C                         IER = 137 - ERROR RETURN FROM C1PNF
C                         IER = 138 - ERROR RETURN FROM DEVAL1
C                         IER = 139 - ABS(DALPHA) IS TOO SMALL.
C                            IF DALPHA = 0, THE PRESENT
C                            FORMULAE ARE NOT APPLICABLE.
C                         IER = 140 - ABS(OMEGA) IS TOO SMALL.
C                            BIFURCATION MAY BE SIMPLE (EV1
C                            REAL), NOT A HOPF BIFURCATION.
C
C   SUBROUTINES REQUIRED -   REF, COPY, CHECKJ, ANUCRT,
C                            IGUESS, EXTRAP, EVALS, EVAL1,
C                            RCCOPY, DEVAL1, NWTN, ENORM2,
C                            INITER, ENRML, C1PNF, BFNRML,
C                            RLNRML, DIF2, PRJCT2, CMAN2,
C                            DIF3, DGFUN, EIGR, UERTST,
C                            *SGEFA, *SGESL, *ISAMAX,
C                            *SAXPY, *SDOT, *SSCAL,
C                            *CGEFA, *CGESL, *CAXPY, *CDOTC,
C                            *CSCAL, *ICAMAX,
C                            **BALANC, **ELMHES, **HQR
C                          * INDICATES LINPACK ROUTINE,
C                         ** INDICATES EISPACK ROUTINE.
C   MACHINE DEPENDENT CONSTANTS -
C                         BALANC CONTAINS RADIX, THE BASE OF
C                            MACHINE FLOATING POINT NUMBERS.
C                         HQR CONTAINS MACHEP, THE RELATIVE
C                            PRECISION OF THE FLOATING POINT
C                            ARITHMETIC.
C
C*****************************************************************
C                                                                *
C   DESCRIPTION OF SUBROUTINE FNAME(X,N,ANU,F,A,IND)             *
C                                                                *
C      THIS SUBROUTINE MUST EVALUATE THE FUNCTION F=F(X;ANU)     *
C      AND THE JACOBIAN MATRIX A=DF/DX(X;ANU) FOR THE            *
C      SPECIFIC SYSTEM OF O.D.E.'S. FNAME MUST BE DECLARED       *
C      EXTERNAL IN THE PROGRAM WHICH CALLS BIFOR2.               *
C                                                                *
C   PARAMETERS          X       - X IS AN N-VECTOR CONTAINING    *
C                                   THE POINT AT WHICH F OR      *
C                                   A IS TO BE EVALUATED.        *
C                       N       - N IS THE DIMENSION OF X.       *
```

```
C                        ANU    - ANU IS THE VALUE OF THE        *
C                                  BIFURCATION PARAMETER FOR     *
C                                  WHICH F OR A IS TO BE         *
C                                  EVALUATED.                    *
C                        F      - F IS AN N-VECTOR. WHEN CALLED  *
C                                  WITH IND=0, FNAME MUST        *
C                                  EVALUATE F(X;ANU) AND         *
C                                  STORE THE RESULT IN F .       *
C                        A      - A IS AN N BY N MATRIX. CALLED  *
C                                  WITH IND=1, FNAME MUST        *
C                                  EVALUATE THE JACOBIAN         *
C                                  AND STORE THE RESULT IN A,    *
C                                  A(I,J) = DF(I)/DX(J),         *
C                                     I,J=1,...,N .              *
C                        IND    - IND INDICATES WHETHER FNAME    *
C                                  IS TO EVALUATE F OR A.        *
C                               IF IND = 0, EVALUATE F,          *
C                               IF IND = 1, EVALUATE A.          *
C                                                                *
C*****************************************************************
C
      DIMENSION XS(N),PAR(10),ERR(7),W(1)
      COMPLEX V1(N)
      COMPLEX EV1,C1,C1ERR,DEV,DEVERR
      EXTERNAL FNAME,EVALS,EVAL1
C
      IF(N .LT. 2) GO TO 9129
      IF(U .LE. 0.0) GO TO 9130
      IF((MTH.LT.1).OR.(MTH.GT.2)) GO TO 9131
      IF((JJOB.LT.0).OR.(JJOB.GT.1)) GO TO 9132
      IF((IPRINT.LT.0).OR.(IPRINT.GT.2)) GO TO 9133
C
C-----------------------------------------------------------------
C                          EPSR IS THE TOLERANCE USED TO TEST
C                             FNAME FOR CONSISTENCY,(CHECKJ).
C                          EPSR IS ALSO USED TO TEST IF
C                             ABS(OMEGA) OR ABS(DALPHA) ARE
C                             TOO SMALL.
C                          EPS, NSIG AND ITMAX CONTROL THE
C                             VARIOUS ITERATIONS.
C                          THE VALUES ASSIGNED BELOW WERE FOUND
C                             SATISFACTORY FOR ALL THE PROBLEMS
C                             CONSIDERED IN H. K. AND W.
      EPSR = 1.0E-4
      EPS = 1.0E-13
      NSIG = 10
      ITMAX = 20
C
C-----------------------------------------------------------------
```

Appendix F. A Sample Program Using BIFOR2

The program listed below used BIFOR2 to analyze the Hopf bifurcation which occurs in the system

$$\dot{x}_1 = x_2$$

$$\dot{x}_2 = (\sin x_1 \cos x_2)\, x_3^2 - \sin x_1 - \gamma x_2$$

$$\dot{x}_3 = \kappa(\cos x_1 - \rho)$$

for the parameter values $\kappa = .1$ and $\rho = .2$, regarding γ as the bifurcation parameter. This system is the model centrifugal governor-steam engine described in the Introduction, Chapter 1, and in Example 2 of Chapter 3. Essentially the same program appears on the microfiche as subroutine CG.

Remark 1. The use of labelled COMMON in this example is to illustrate how additional parameters in the system may be passed to the user-supplied subroutine FNAME, which in this case is subroutine CGFUN.

Remark 2. The bifurcation parameter in this example is called GAM both in the main program and in subroutine CGFUN. In output written from BIFOR2, however, the bifurcation parameter is always referred to as ANU.

Remark 3. In this example, the critical value of the bifurcation parameter GAM and the corresponding stationary point XS happen to be known analytically. The example thus does not illustrate the ability of BIFOR2 to compute these values.

Remark 4. The number u is machine dependent. The value $u = 1.0\text{E-}14$ is appropriate for a machine in which the floating point arithmetic carries roughly 14 decimal digits.

PROGRAM LISTING

```
      DIMENSION XS(3),PAR(10),ERR(7),W(54)
      COMPLEX V1(3)
C                                  DIMENSION OF W = 2*N**2 + 12*N
      COMMON/CGC/AKAPPA,RHO
      EXTERNAL CGFUN
      AKAPPA = .1
      RHO = .2
      XS(1) = ACOS(RHO)
      XS(2) = 0.0
      XS(3) = SQRT(1.0/RHO)
      N = 3
      GAM = 2.0*AKAPPA*(RHO**(1.5))
      U = 1.0E-14
      MTH = 1
      JJOB = 1
      IPRINT = 2
      CALL BIFOR2(CGFUN,XS,N,GAM,U,MTH,JJOB,IPRINT,
     1            PAR,V1,ERR,W,IER)
      STOP
      END
      SUBROUTINE CGFUN(X,N,GAM,F,DF,IND)
C                                  EVALUATE FUNCTION AND JACOBIAN FOR
C                                  THE CENTRIFUGAL GOVERNOR-STEAM
                                   ENGINE
      DIMENSION X(N),F(N),DF(N,N)
      COMMON/CGC/AKAPPA,RHO
      SINX1 = SIN(X(1))
      COSX1 = COS(X(1))
      IF(IND .EQ. 1) GO TO 5
C                                  EVALUATE FUNCTION
      F(1) = X(2)
      F(2) = (SINX1*COSX1)*X(3)**2 - SINX1 - GAM*X(2)
      F(3) = AKAPPA*(COSX1 - RHO)
      RETURN
C                                  EVALUATE JACOBIAN
    5 DF(1,1) = 0.0
      DF(1,2) = 1.0
      DF(1,3) = 0.0
      DF(2,1) = (COSX1**2-SINX1**2)*X(3)**2 - COSX1
      DF(2,2) = -GAM
      DF(2,3) = 2.0*SINX1*COSX1*X(3)
      DF(3,1) = -AKAPPA*SINX1
      DF(3,2) = 0.0
      DF(3,3) = 0.0
      RETURN
      END
```

This program will produce one entry of Table 3.2, p. 154.

Appendix G. Contents of Microfiche

The microfiche contains

i) an index for the microfiche

ii) a driver program

iii) application subprograms

iv) BIFOR2 and associated subprograms

v) output produced by executing the entire program .

The job represented by the microfiche was run on a CDC Cyber 174, using the FTN 4.8 compiler, OPT = 2. Compilation and execution each took slightly less than one minute of CPU time.

The driver program merely calls the application subprograms. The application subprograms are divided into groups, the first two letters of the names within a group indicating the application. If a subroutine name ends with 'FUN', then that subroutine evaluates the function f and the Jacobian $\partial f/\partial x$ for a specific application. The suffix 'XCT' indicates that the subroutine evaluates exact bifurcation formulae. The suffix 'TAB' indicates that the subroutine creates table(s).

The prefixes indicating applications are as follows:

SB- mass-spring belt problem (Chapter 1, Introduction,
Chapter 2, Example 1,
Chapter 3, Example 1)

VP- van der Pol equation (Chapter 2, Example 2)

BR- Brusselator (Chapter 2, Example 3)

LN- Langford system (Chapter 2, Example 4)

TR- Tori generating system (Chapter 2, Example 4)

PP- Predator prey equation (Chapter 2, Exercise 5)

FN- Fitzhugh-Nagumo system (Chapter 2, Exercise 6)

HD- Howard system (Chapter 2, Exercise 7)

CG- Centrifugal governor-steam engine (Chapter 1, Introduction,
Chapter 3, Example 2)

.R- Lorenz system (Chapter 3, Example 3)

IH- Hodgkin-Huxley system (Chapter 3, Example 4)

BD- Brusselator with fixed b.c.'s (Chapter 3, Example 5,
Chapter 5, Section 5)

'F- Panel flutter problem (Chapter 3, Example 6)

:B- Kubiček system (Chapter 3, Exercise 3)

IW- Hutchinson-Wright equation (Chapter 4, Section 3,
Chapter 4, Exercise 5)

Y- Delay system studied by Kaplan-Yorke (Chapter 4, Section 4,
Chapter 4, Exercise 5)

IA- May's three-trophic level delay system (Chapter 4, Section 6)

Note that these same prefixes may be used to locate the output from the various application subprograms; see the index on he microfiche.

REFERENCES

[1] Abraham, R. and Robbin, J.
'Transversal Mappings and Flows', W. A. Benjamin Inc., N. Y. (1967)

[2] Airy, G. B.
On the regulation of the clockwork for effective uniform movement of equatorials, Mem. Roy. Astron. Soc., vol. 1 (1840), 249-287, see also vol. 20 (1851), 115-119

[3] Allwright, D. J.
Harmonic balance and the Hopf bifurcation, Math. Proc. Camb. Phil. Soc. 82 (1977), 453-467

[4] Andronov, A. A., Vitt, A. A. and Chaikin, S. E.
'Theory of Oscillations', 2nd ed., Moscow, State Press for Phys.-Math. Lit., 915p. (1959), in Russian, English translation, Pergamon Press N. Y. (1966)

[5] Arnol'd, V. I.
'Ordinary Differential Equations', Translated and edited by Richard A. Silverman, MIT Press, Cambridge Mass. (1973)

[6] Arnol'd, V. I.
'Supplementary Chapters of the Theory of Ordinary Differential Equations', Nauka, Moscow (1978), 304 pp. (in Russian)

[7] Auchmuty, J. F. G. and Nicolis, G.
Bifurcation analysis of reaction-diffusion equations III, Chemical Oscillations, Bull. Math. Biology 38 (1976), 325-350

[8] Balakrishnan, A. V.
'Applied Functional Analysis', Springer-Verlag (1976)

[9] Beddington, J. R. and May, R. M.
Time delays are not necessarily destabilizing, Math. Biosci. 27 (1975), 109-117

[10] Birkhoff, G. D.
'Dynamical Systems', A. M. S. Colloquium Publications vol. IX, Providence, R. I. (1927), revised edition (1966)

[11] Boa, J. A. and Cohen, D. S.
Bifurcation of localized disturbances in a model biochemical reaction, SIAM J. Appl. Math. 30 (1976) 123-135

[12] Bochner, S. and Montgomery, D.
Groups of differentiable and real or complex analytic transformations, Ann. of Math. 46 (1945), 685-694

[13] Boyce, W. E. and DiPrima, R. C.
'Elementary Differential Equations and Boundary Value Problems', 3rd ed. John Wiley and Sons, N. Y. (1977) xiv + 582 pp

[14] Buck, R. C.
'Advanced Calculus', McGraw-Hill, N. Y. (1956) VIII + 423 pp

[15] Chafee, N.
The bifurcation of one or more closed orbits from an equilibrium point of an autonomous differential system, J. Diff. Eqns. 4 (1968), 661-679

[16] Chafee, N.
A bifurcation problem for a functional differential equation of finitely retarded type, J. Math. Anal. and Applics. 35 (1971), 312-348

[17] Chernoff, P. R. and Marsden, J. E.
'Properties of Infinite Dimensional Hamiltonian Systems', Lecture Notes in Mathematics vol. 425, Springer-Verlag, N. Y. (1974)

[18] Chow, S. N. and Mallet-Paret, J.
Integral averaging and Hopf bifurcation, J. Diff. Eqns. 26 (1977), 112-159

[19] Claeyssen, J. R.
Effect of delays on functional differential equations, J. Diff. Eqns. 20 (1976), 404-440

[20] Coddington, E. A. and Levinson, N.
'Theory of Ordinary Differential Equations', McGraw-Hill, N. Y. (1955) 429 pp

[21] Conte, S. D. and de Boor, C.
'Elementary Numerical Analysis', 3rd edition, McGraw-Hill, N. Y. (1980)

[22] Cooley, J. W. and Dodge, F. A.
Digital computer solutions for excitation and propogation of nerve impulse, Biophys. J. 6 (1966), 583-599

[23] Copson, E. T.
'An Introduction to the Theory of Functions of a Complex Variable', Oxford at the Clarendon Press (1935)

[24] Courant, R. and Hilbert, D.
'Methods of Mathematical Physics', vols. I, II, Interscience, N. Y. (1953 and 1962)

[25] Crandall, M. G. and Rabinowitz, P. H.
Bifurcation from simple eigenvalues, J. Funct. Anal. 8 (1971), 321-340

[26] Crandall, M. G. and Rabinowitz, P. H.
The Hopf bifurcation theorem in infinite dimensions, Arch. Rat. Mech and Anal. 67, 1 (1977), 53-72

[27] Dahlquist, G. and Björck, A.
'Numerical Methods', Prentice Hall, Englewood Cliffs, N. J. (1974)

[28] Decker, D. W. and Keller, H. B.
Solution branching - a constructive approach, in 'New Approaches in Nonlinear Dynamics', Holmes, P. and Othmer, H. (eds.), SIAM publications, Philadelphia (1980)

[29] Dongarra, J. J., Moler, C. B., Bunch, J. R. and Stewart, G.
'LINPACK Users' Guide', SIAM publications, Philadelphia (1979)

[30] Dowell, E. H.
'Aeroelasticity of Plates and Shells', Sitzhoff and Noordhoff Int. Publ., Winchester, Mass. (1975)

[31] Fife, P. C.
Asymptotic states for equations of reaction and diffusion, Bull. Amer. Math. Soc. 84 (1978) 693-726

[32] Fife, P. C.
'Mathematical Aspects of Reacting and Diffusing Systems', Lecture Notes in Biomathematics, vol. 28, Springer-Verlag N. Y. (1979)

[33] Floquet, G.
Sur les équations différentielles lineáres à coefficients périodiques, Ann. École Norm. Sup. Paris (2) 12 (1883), 47-89

[34] Friedman, A.
'Partial Differential Equations', Holt, Rinehart and Winston, N. Y. (1969)

[35] Frisch, R. and Holme, H.
The characteristic solutions of a mixed difference and differential equation occurring in economic dynamics, Econometrica 3 (1935), 225-239

[36] Garabedian, P. R.
'Partial Differential Equations', John Wiley and Sons, N. Y. (1964) XII + 672 pp

[37] Grafton, R. B.
A periodicity theorem for autonomous functional differential equations, J. Diff. Eqns. 6 (1969), 87-109

[38] Halanay, A.
'Differential Equations; Stability, Oscillations, Time Lags', Math. in Sci. and Eng. vol. 23, Academic Press, N. Y. (1966)

[39] Hale, J. K.
'Ordinary Differential Equations', Wiley-Interscience, N. Y. (1969)

[40] Hale, J. K.
'Functional Differential Equations', Applied Math. Sci. vol. 3, Springer-Verlag, N. Y. (1971)

[41] Hale, J. K.
Nonlinear oscillations in equations with delays, 'Nonlinear Oscillations in Biology', F. C. Hoppensteadt. ed., Lectures in Applied Mathematics, Vol. 17, Amer. Math. Soc., Providence, R. I. (1979)

[42] Hale, J. K. and Kato, J.
Phase space for retarded equations with infinite delays, Funk. Ekvacioj 21 (1978), 297-315

[43] Hartman, P.
'Ordinary Differential Equations', Baltimore (1973) XIV + 612 pp

[44] Hassard, B. D.
Bifurcation of periodic solutions of the Hodgkin-Huxley model for the squid giant axon, J. Theoretical Biology 71 (1978), 401-420

[45] Hassard, B.
The numerical evaluation of Hopf bifurcation formulae, 'Information Linkage between Applied Mathematics and Industry', P. C. C. Wang (ed.), Academic Press, N. Y. (1979)

[46] Hassard, B. D. and Wan, Y. H.
Bifurcation formulae derived from center manifold theory, J. Math. Anal. and Applics. 63 (1978), 297-312

[47] Hausrath, A. R.
Stability in the critical case of purely imaginary roots for neutral functional differential equations, J. Diff. Eqns. 13 (1973), 329-357

[48] Henry, D.
'Geometric Theory of Semilinear Parabolic Equations', Mimeographed notes, Univ. of Kentucky (1974)

[49] Hille, E. and Phillips, R. S.
'Functional Analysis and Semigroups', Amer. Math. Soc. Colloq. Publ. vol. 31, Providence, R. I. (1957)

[50] Hirsch, M., Pugh, C. and Shub, M.
'Invariant Manifolds', Lecture Notes in Mathematics vol. 583, Springer-Verlag, N. Y. (1977)

[51] Hirsch, M. W. and Smale, S.
'Differential Equations, Dynamical Systems and Linear Algebra', Academic Press, N. Y. (1974)

[52] Hodgkin, A. L. and Huxley, A. F.
A quantitative description of membrane current and its application to conduction and excitation in nerve, J. Physiol. 117 (1952), 500-544

[53] Holmes, P. and Marsden, J.
Bifurcation to divergence and flutter in flow-induced scillations: an infinite dimensional analysis, (Inst. of Sound and Vibration Research, Univ. of Southhampton), Control of Distributed Parameter Systems, Coventry, England, 28 June-1 July 1977, (Oxford, England; Pergamon, 1978), pp. 133-145

[54] Holmes, P. J. and Rand, D. A.
Identification of vibrating systems by generic modelling, Inst. of Sound and Vibration Research Tech. Rpt. 79, Southhampton (1975)

[55] Hopf, E.
Abzweigung einer periodischen Losung von einer stationaren Losung eines Differentialsystems, Ber. Verh. Sachs. Akad. Wiss. Leipsig Math.-Nat. 94 (1942), 3-22, Translation to English with commentary by L. Howard and N. Kopell, in [81;163-205]

[56] Hopf, E.
A mathematical example displaying the features of turbulence, Comm. Pure. Appl. Math. 1 (1948), 303-322

[57] Hsü, I. D. and Kazarinoff, N. D.
An applicable Hopf bifurcation formula and instablity of small periodic solutions of the Field-Noyes model, J. Math. Anal. and Applics. 55 (1976), 61-89

[58] Hutchinson, G. E.
Circular causal systems in ecology, Ann. N. Y. Acad. Sci. 50 (1948), 221-246

[59] Iooss, G.
Existence et stabilité de la solution périodique secondaire intervenant dans les problèmes d'évolution du type Navier-Stokes, Arch. Rat. Mech. and Anal. 49 (1972), 301-329

[60] Iooss, G.
Bifurcation of a periodic solution of the Navier-Stokes equations into an invariant torus, Arch. Rat. Mech. and Anal. 58, 1 (1975), 34-56

[61] Iooss, G.
Lecture Notes, University of Minnesota (1978)

[62] Ize, J.
'Bifurcation theory for Fredholm operators', Amer. Math. Soc. Memoir No. 174 (1976)

[63] Jack, J. J. B., Noble, D. and Tsien, R. W.
'Electric current flow in excitable cells', Clarendon Press, Oxford (1975)

[64] Jones, G. S.
The existence of periodic solutions of $f'(x) = -a f(x-1)(1 + f(x))$, J. Math. Anal. and Applics. 5 (1962), 435-450

[65] Jones, G. S.
Periodic motions in Banach space and applications to functional differential equations, Contrib. Diff. Eqns. 3 (1964), 75-106

[66] Joseph, D. D. and Sattinger, D. H.
Bifurcating time periodic solutions and their stability, Arch. Rat. Mech. and Anal. 45 (1972), 79-109

[67] Judovich, V. I.
The birth of proper oscillations in a fluid, Prikl. Math. Mek. 35 (1971), 638-655

[68] Kaplan, J. L. and Yorke, J. A.
Ordinary differential equations which yield periodic solutions of differential delay equations, J. Math. Anal. and Applics. 48 (1974), 317-324

[69] Kato, T.
'Perturbation theory for Linear Operators', Springer-Verlag, N. Y. (1966)

[70] Kazarinoff, N. D. and van den Driessche, P.
A model predator prey system with functional response, Math. Biosciences 39 (1978), 125-134

[71] Kazarinoff, N. D., van den Driessche, P. and Wan, Y. H.
Hopf bifurcation and stability of periodic solutions of differential-difference and integro-differential equations, J. Inst. of Math. and Its Applics. 21 (1978), 461-477

[72] Kelley, A.
The stable, center-stable, center, center-unstable, and stable manifolds, J. Diff. Eqns. 3 (1967), 546-570, see also Appendix C of [1]

[73] Kubiček, M.
Algorithm for evaluation of complex bifurcation points in ordinary differential equations, SIAM J. Appl. Math. 38 (1980), 103-107

[74] Lanford, O. E.
Bifurcation of periodic solutions into invariant tori: the work of Ruelle and Takens, 'Nonlinear Problems in the Physical Sciences and Biology', Springer Lecture Notes vol. 322 (1973)

[75] Lefever, R. and Prigogine, I.
Symmetry-breaking instabilities in dissipative systems II, J. Chem. Phys. 48 (1968), 1695-1700

[76] Lima, P.
Hopf bifurcation in equations with infinite delays, Ph. D. Thesis, Brown University, June 1977

[77] Macdonald, M.
Time delay in prey-predator models, Math. Biosciences 28 (1976), 321-330

[78] MacFarlane, A. G. J.
The development of frequency-response methods in automatic control, I.E.E.E. Trans. on Automatic Control, AC-24 (1979), 250-265

[79] Mackey, M. and Glass, L.
Oscillations and chaos in physiological control systems, Science 197 (1977), 287-289

[80] Marsden, J. E.
The Hopf bifurcation for nonlinear semigroups, Bull. Am. Math. Soc. 79 (1973), 537-541

[81] Marsden, J. E. and McCracken, M.
'The Hopf Bifurcation and Its Applications', Applied Math. Sciences, vol. 19, Springer-Verlag, N. Y. (1976) XIII + 408 pp

[82] Massera, J. L.
On Lyapunov's conditions of stability, Annals of Mathematics (2), 50 (1949), 705-721

[83] Maxwell, J. C.
On governors, Proc. Roy. Soc. London, vol. 16 (1868), 270-283

[84] May, R.
Time-delay versus stability in population models with two and three trophic levels, Ecology 54 (1973), 315-325

[85] Mizohata, S.
'The Theory of Partial Differential Equations', Cambridge U. Press (1973)

[86] Morris, H. C.
A perturbative approach to periodic solutions, of delay-differential equations, J. Inst. of Math. and Applics. 18 (1976), 15-24

[87] Naito, T.
On autonomous linear functional differential equations with infinite retardations, J. Diff. Eqns. 21 (1976), 297-315

[88] Negrini, P. and Salvadori, L.
Attractivity and Hopf Bifurcation, Nonlinear Analysis, Theo. and Appl. 3 (1979), 87-99

[89] Ortega, J. M. and Rheinboldt, W. C.
'Iterative Solution of Nonlinear Equations in Several Variables', Academic Press, N. Y. (1970)

[90] Palais, R.
'Foundations of Global Non-linear Analysis', W. A. Benjamin, Inc., N. Y. (1968)

[91] Peters, G. and Wilkinson, J. H.
Inverse iteration, ill-conditioned equations and Newton's method,SIAM Review 21 (1979), 339-360

[92] Poincaré, H.
Sur les courbes définies par une équation différentielle, J. Math. Pures Appl. (4) 1 (1885), 167-244

[93] Pontryagin, L. S.
'Ordinary Differential Equations', Addison-Wesley, Reading, Mass. (1962) VI + 298 pp

[94] Poore, A. B.
On the theory and application of the Hopf-Friedrichs bifurcation theory, Arch. Rat. Mech. and Anal. 60 (1975-76), 371-393

[95] Rellich, F.
Ein Satz über mittlere Konvergenz, Nachr. Akad. Wiss. Gottingen, Math.-Phys. Kl. 30-35 (1930)

[96] Rinzel, J. and Miller, J. N.
Numerical calculation of stable and unstable periodic solutions to the Hodgkin-Huxley equations, to appear, Math. Biosciences (1980)

[97] Ruelle, D. and Takens, F.
On the nature of turbulence, Comm. Math. Phys. 20 (1971), 167-192 and 23 (1971), 343-344

[98] Ruppelt, R. and Schneider, A.
Über ein numerishes verfahren zur bestimmung der qualitativen eigenshaften bei ber Hopf-bifurkation, Diplomarbeit, Universität Bremen (1979)

[99] Segal, I.
Nonlinear semigroups, Ann. of Math. 78 (1963), 339-364

[100] Smith, B. T., Boyle, J. M., Garbow, B. S., Klema, V. C. and Moler, C. B., 'Matrix Eigensystem Routines', Springer-Verlag (1974)

[101] Stetch, H. W.
On the adjoint theory for autonomous linear functional differential equations, J. Diff. Eqns. 27 (1978), 421-443

[102] Stewart, W. E., Ray, W. H. and Conley, C. C. (editors) 'Dynamics and Modelling of Reactive Systems', Academic Press, (1980), Proceedings of Advanced Symposium on Dynamics and Modelling of Reactive Systems, Mathematics Research Center, Madison, Wisconsin, Oct 22-24, 1979.

[103] Stirzaker, D.
On a population model, Math. Biosciences 23 (1975), 329-336

[104] Van Gils, S.
Hopf bifurcation and attractivity, Stichting Mathematisch Centrum, Amsterdam, TN93 (1979)

[105] Van Strien, S. J.
Center manifolds are not C^{∞}, Mathematische Zeitschrift 166 (1979), 143-145

[106] Vyshnegradskii, I. A.
Sur la theorie des regulateurs, C. R. Acad. Sci. Paris 83 (1876), 318-321

[107] Walter, G. G.
J. Fisheries Res. Board Canada 30 (1973), 939-945

[108] Wan, Y. H.
On the uniqueness of invariant manifolds, J. Diff. Eqns. 24 (1977), 268-273

[109] Wilkinson, J. H.
'Rounding Errors in Algebraic Processes', H. M. Stationery Office, London (1963)

[110] Wilkinson, J. H.
'The Algebraic Eigenvalue Problem', Clarendon Press, Oxford (1965)

[111] Wright, E. M.
A nonlinear difference-differential equation, J. Reine und Angew. Math. 194 (1955), 66-87

[112] Young, D. M. and Gregory, R. T.
'A Survey of Numerical Mathematics, Volume I', Addison-Wesley, Reading, Mass. (1972)

INDEX